ISBN-13: 978-1530091478
ISBN-10: 1530091470
Original publication, October 5, 2015
Edition 2, February 20, 2016
LCCN: 2016903336
Publisher: CreateSpace
North Charleston, SC

GNU Octave Primer for Beginners
EZ Guide to the Commands and Graphics
Written for GNU OCTAVE-4.0.0

Second Edition

S. Nakamura

Printed by
CreateSpace
7290 Investment Dr, Suite B
North Charleston, SC 29418, USA

About the image on the front cover

A fractal image based on Henon's model is plotted using the Octave's graphic functions on a spherical surface, creating a virtual planet. Reference: H. Lauerier, *Fractals*, Princeton University Press, 1991. The original equations are:

$$X_{n+1} = 0.24x_n - 0.9708(y_n - x_n^2)$$
$$Y_{n+1} = 0.9708x_n + 0.24(y_n - x_n^2)$$

Table of Contents

Preface

This book serves as a hands-on tutorial for beginners who are unfamiliar with Octave. This is the second edition of the same title with many enhancements including added material in existing sections, additional sections and subsections, and exercise problems with full solutions. Lists of a large number of programs used to solve exercise problems may be helpful to the readers also. Print version and Kindle version are both available.

Octave (also called GNU Octave) is a high-level programming language useful in science, engineering and business. It is a great tool in studying and practicing mathematics and data processing. Octave is compatible with Matlab developed by Math Works.

Yet one significant difference is that Octave is available free of any charge. Octave is available for Windows, Linux, and OS X. Although we use PC with *Windows* in this writing, all written in this book applies to other systems except for minor differences in file management procedures. So please do not be discouraged to read further even if you use other computer system than Windows.

The first thing to do is to download Octave, which is free of any charges, from the following site:
> https://www.gnu.org/software/octave/download.html

Octave is very sophisticated math software, yet it can also be used as a simple calculator on your PC or Mac or Linux computer. But unlike any ordinary calculator, Octave is a programmable calculator and has useful graphic tools which are very important in studying algebra and calculus.

Octave needs a little tutorial to begin with. That is the purpose of this writing.

Scripts (programs) are exchangeable between Octave and MAT-LAB although some minor adjustments may be necessary. Such

4

adjustments are no greater than what can happen even between two versions of Matlab.

Octave is said to be suitable for numerical methods. That is true, but almost all needs in mathematical computations in science and engineering can be fulfilled with Octave whether the numerical methods are used or not. Octave is also very useful in analyzing and processing financial data. Many users of Fortran, C and C++ find the programming capabilities of Octave more useful than Fortran, C and/or C++ because of high quality of computations, easy programming, convenient tools for various chores, and superior graphic capabilities.

The only major drawback of Octave is that its processing speed is significantly slower than Fortran, C, or other similar programming languages. However, this is not a great problem unless very lengthy computations such as solving partial differential equations are done. For such heavily computational project, Fortran, C or other similar language should be used, but yet the results can be nicely post-processed by Octave.

For general background of Octave, read the introductory article on Wikipedia: https://en.wikipedia.org/wiki/GNU_Octave

To those who want to learn more of the capabilities and potential applications, the books and articles written for Matlab should be helpful, because what can be done with Matlab can be done with Octave, and vice versa. Some of references are listed at the end of this book.

Finally, note that the developers of Octave request donations from those who find Octave very useful. This request is well justified because the developers spent their own time and money to develop Octave and refine it.

S. Nakamura
Author

Chapter 1
Octave Commands and Programming

1.1 Before starting calculations

After download is completed, Octave-4.0.0-installer.exe can be found in *Download* directory. To install Octave, click on it, and you will be guided by the installation wizard. When installation is finished, icons of Octave-4.0.0(CLI) and Octave-4.0.0(GUI) shown below will appear on the desktop.

Clicking on the former will open CLI, meaning Command Line Interface, and the latter Graphic User Interface. We use the former first. Once you learn with CLI, switching to the other is easy as explained later. If you prefer to start with GUI, that is fine too, because GUI has also a command window that is equivalent to CLI. One advantage with the command window of GUI is that copy & paste operation is allowed there, while copy & past is not possible in CLI.

The CLI window is illustrated in Figure 1. The **>>** sign is called *Command Prompt*, after which a small white box is flashing. You can write a command there. Type **date** as the first trial and hit *Return*. Then, today's date is printed like:

ans = 17-Sep-2015

Likewise if you type **fix(clock)**, the response is like

ans = 2015   9   17   10   50   52

which are year, month, day, hour, minute, second, respectively. Also try simply **clock** without writing **fix** to see what happens.

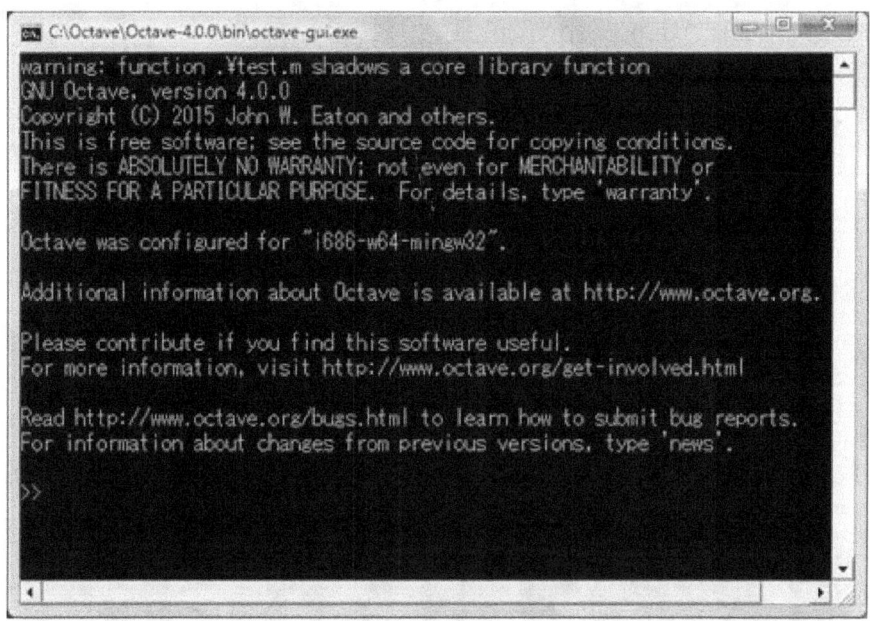

Figure 1.1 CLI

The most important thing at this stage is to know in what directory (or folder) you are in. To do this, type **pwd** after the prompt. A typical response is

    ans = c:/users/owner

which means currently the directory is **owner** in **users** in **c** drive. Here, **owner** is used as an example name, so the directory name owner may be different for each PC.

To know what files and directories (or folders) are in the current directory, type **ls** or **dir** after the prompt.

To change the current directory to one of other directories (or folders) contained in the current directory, for example, **sf1**, type **cd sf1** after the prompt >>. Here you can type **pwd** to make sure the directory has been changed. If the answer is

ans = c:/users/owner/sf1

you are sure that the current directory is **sf1** in **owner** in **users** in **c** drive. To go back to the parent directory **owner**, type **cd ..** (one blank space followed by dot dot). If you want to go to directory **users** while you are in **sf1,** type **cd ../...**   Likewise, to go to **c** drive from **sf1**, type **cd ../../...**

To change directory to another drive like **h** drive (say a stick memory), type **cd h:**. If you have changed to drive **h** but want to come back to **owner** in **users** in drive **c**, type **cd c:/ users/owner**.

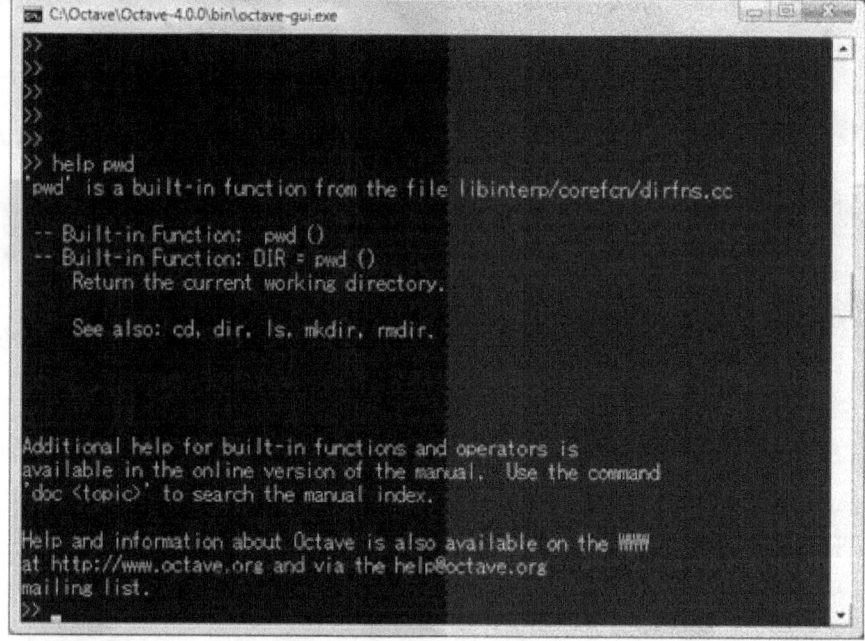

Figure 1.2   Response to **help pwd**

To get some information about a command, use **help**. For example, to get information about the **pwd** command, type **help pwd** after the prompt. The response is illustrated in Figure 1.2.

8

Since this image of the window is fuzzy, you should do it by yourself and read on your own computer.

A useful hint: In revising the statement while working on CLI, you do not have to retype the whole things. Use the arrow keys like ↑ and ↓. The lines written earlier will appear, each of which can be revised by first erasing using the *Backspace* key and then rewriting that part and hit *Return*. Some important commands you cannot bypass are explained next.

mkdir

mkdir creates a directory. >> **mkdir folder_a** creates a directory named **folder_a** in the current directory.

When you work with Octave, you are likely to create many files in the directory. Working in a high level directory such as **users** or **owner** is not desirable. Therefore, it is recommended to make a working directory for each project or period of time by **mkdir** and work in that. Assuming your working directory is to be named **MyOctavePlace,** use **mkdir** as follows:

>>mkdir MyOctavePlace

Then, change directory by

>>cd MyOctavePlace

Assuming **MyOctavePlace** was created in **owner**, make sure that **MyOctavePlace** is in **owner** by **pwd**:

ans = c:/users/owner/ MyOctavePlace

Of course, **MyOctavePlace** can be created in a different (parent) directory if you choose so.

Because **c:/users/owner** is the default (or current) directory when Octave is started, you have to change directory manually by >>**cd MyOctavePlace** each time.

However **MyOctavePlace** can be made Home directory (or default directory) by using **setenv** command. Detail of this procedure is explained in Appendix 3. It is suggested for the beginners to ignore this setting until he/she becomes more familiar to the Octave operations.

delete
Command **delete** deletes a file but not a directory. >> **delete file_a** deletes the file named **file_a**.

rmdir
Command **rmdir directoryname** deletes a directory named **directoryname**. However it does not work unless the directory is empty. Therefore, if you wish to delete a directory, you have to delete everything in the directory first.

quit
The **quit** command will close the CLI window, and GUI window as well.

diary on, diary off
With **diary** command, you can print all activities on the command window. The **diary filename** command starts writing all keyboard input and screen output to the file named **filename**. Command **diary off** terminates writing. Once **diary filename** is used, you don't have to write it after **diary**: just write **diary** only. All the recording is appended in the **filename** file, which was used in the prior diary writing. The file may be opened for reading in the Editor of GUI. If a suffix **txt** is added to the file name, like **filename.txt**, thc file may be opened in *Notepad*. Files named like **filename.doc** may be opened in MS Word.

1.2 Using Octave as a powerful calculator

Octave can be used as a handy calculator, although it is significantly more powerful than any ordinary calculator. Using Octave as a calculator is a good start in learning Octave.

Arithmetic operations

Arithmetic operators +, -, *, and / are, respectively, addition, subtraction, multiplication, and division operators, just as in traditional programming languages. In addition, ^ is the power operator.

Octave has one untraditional operator \ which may be named *inverse division.* This operator yields the reciprocal of division, that is, **a\b** equals **b/a**. For example,

>> c = 3\1
c = 0.3333

Do not use this operator in usual computations, but it will become useful if you do linear algebra with Octave.

Calculations with scalar variables

As a simple example, let us evaluate the volume of a sphere with radius r = 2:

$$\text{Volume} = 4\pi r^3/3 \quad \text{with} \quad r = 2$$

The commands to type are:

>> r = 2;
>> vol = 4*pi*r^3/3;

where **pi** is the name of π, that is, 3.14159265358979 in Octave. The caret symbol ^ after **r** is the power operator. The first command **r = 2** is to define the value of **r**, which Octave memorizes. Notice in the preceding script that each line is terminated by a semicolon. The operator ; is to tell Octave to be quiet, that is instructing not to show the results of the command or computation immediately.

When we work in the command window, the computer calculates the answer for each command immediately after the return key is hit. Therefore, the value of **vol** is already in the computer. How can we get the result printed out on the screen?

The quickest way of printing out the result is to type

>> vol

Then, the response is

vol = 33.5103216382911

Of course we could have written as

>> r = 2;
>> vol = 4*pi*r^3/3

where the second line is not terminated by the semicolon. Then, the response is immediately

vol = 33.5103216382911

This number is very long because all the numbers in Octave are in double precision. If you want a shorter number, you can tell Octave to print numbers in a short format by

>> format short

After this, the Octave response becomes in the short format like

vol = 33.510

To revert to the long format, the command is

>> format long

If you wish to delete variables like **r** or **vol**, use the command **clear**:

>> clear r
>> clear vol

or

>> clear r vol

Multiple commands (or definitions) can be written in one line, for example,

>> a=1; b=2; c=4;

If each is separated by a comma like

>> a=1, b=2, c=4

then, Octave responds as

a = 1
b = 2
c = 4

Although we did not use the arithmetic operators such as +, -, *, / and ^ yet, they can be used as needed. For example, $y = x^2 + 3x - 5$ with x=3 is calculated as

>> x=3; y=x^2 + 3*x -5
y = 13

The second line above is the answer.

1.3 Calculations involving mathematical functions

Octave has a large number of mathematical functions. Some elementary mathematical functions available in Octave are shown in TABLE 1.1.

For more advanced mathematical functions such as Bessel function or Legendre functions, detailed information is available by using the commands

>> help bessel

or

>> help legendre

Table 1.1 Elementary Functions

Trigonometric functions	Remarks
sin(x) cos(x) tan(x) asin(x) acos(x) atan(x) atan2(y, x) sinh(x) cosh(x) tanh(x) asinh(x) acosh(x) atanh(x)	$\pi/2 \geq atan(x) \geq -\pi/2$ Same as atan(y/x) but $\pi \geq atan2(x,y) \geq -\pi$
Other elementary functions	Remarks

abs(x)
angle(x)
 	
sqrt(x)
real(x)
imag(x)
conj(x)
round(x)
fix(x)
floor (x)
ceil(x)
sign(x)
mod (x, y)
	Absolute value of x
	Phase angle of complex value:
	    if x = real, angle = 0
	    if x = complex, $-\pi/2 <$ angle $< \pi/2$
	Square root of x
	Real part of complex value x
	Imaginary part of complex value x
	Complex conjugate of x
	Round to the nearest integer
	Round a real value x toward zero
	Round a real value x toward $-\infty$
	Round a real value x toward $+\infty$
	+1 if x > 0; -1 if x < 0
	Remainder upon division: x - y*fix(x/y)

rem (x, y)	Remainder upon division: x - y*fix(x/y): different from mod if y $\leq$ 0
exp(x)	Exponential base e
log(x)	Log base e
log10(x)	Log base 10
factor(x)	Factorizes x into prime numbers
isprime(x)	1 if x is a prime number, 0 if not
factorial(n)	(n)(n-1)(n-2)...(3)(2)(1) if n is a positive integer
sort(x)	Rearrange the numbers in array x to increasing order
sum(x)	Computes the total of the numbers in array x
min(x)	Finds the minim in x
max(x)	Finds the maximum in x
rand('seed', x)	Sets the seed number to x
rand	Generates a random number between 0 and 1
rand(n)	n-by-n matrix of random numbers

The mathematical functions can be used in any calculations for example,

>> sin(1.2)*exp(1)

which returns the value of sin(1.2)exp(1), namely, **ans = 2.5335**. The trigonometric functions use radian only. So if a value is in degree, you have to convert it before using the trigonometric functions or in the arguments in the trigonometric functions like

>> sin(angle_1*pi/180)*exp(named_variable)

1.4 Variables and variable names

Variable names and their types do not have to be declared. This is because variable names in Octave make no distinction among integer, real, and complex variables. Any variable can take real, complex, and integer values in double precision.

In principle, any name can be used as long as it is compatible in Octave. We should, however, be aware of two incompatible situations. The first is that the name is not accepted by Octave. The second is that the name is accepted, but it destroys the original meaning of a reserved name. These conflicts can occur with the following types of names:

(a) Names for certain values
(b) Function (subroutine) names
(c) Command names

A bad example of the second conflict is as follows: If **sin** and **cos** are used as user-defined variables, for example,

>> sin = 3;
>> cos = sin - 2;

the calculations proceed; however, **sin** and **cos** can never be used as trigonometric functions thereafter until variables are cleared by **clear** or Octave is shut down.

Traditionally, symbols i, j, k, 1, m, and n have been used as integer variables or indices. At the same time, i and j are used to denote a unit imaginary value, $\sqrt{(-1)}$. In Octave, **i** and **j** are reserved as unit imaginary value. Therefore, if the computation involves complex variables, it is advisable to avoid **i** and **j** as user-defined variables or indices. (Because i and j are so commonly used as indices in mathematics, it is hard not to use these as indices at all. The author uses these as indices in programs which never involve imaginary or complex values.)

Table 1.2 lists reserved variable names that have special meanings.

In order to examine if a variable or function name you consider is in any conflict, use **exist**. For example, for >>**exist atan**, the response is

ans = 5

where **5** means that **atan** is a built-in function, so it cannot be used as your user-defined name.

In general, **>>exist name** is responded as one of the following cases:

if **ans=1**, **name** is a variable;

if **ans=2**, **name** is an absolute file name, an ordinary file in Octave's 'path', or (after appending '.m') a function file in Octave's 'path';

if **ans=3**, **name** is a '.oct' or '.mex' file in Octave's 'path';

if **ans=5**, **name** is a built-in function;

if **ans=7**, **name** is a directory;

if **ans=103**, **name** is a function not associated with a file (entered on the command line);

if **ans=0**, **name** does not exist.

Therefore, if **ans=0,** the name can be safely used as a variable or file name or function name. Special numbers and variable names are shown in Table 1.2.

Table 1.2   Special Numbers and Variable Names

```
epsilon: machine epsilon= 2.22204e-16
pi: π = 3.141592653589793
i and j: unit imaginary = √(-1)
inf: infinity, ∞
nan: not a number
date: date
clock: clock
flops: floating point operation count
nargin: number of function input count
nargout: same for function output
```

String variables

Any compatible variable names can be used to represent a string. For example,

```
>> s ='My name is Tom Brown'
s = My name is Tom Brown
```

Here, the variable **s** is a string variable, which is an array of characters. More details of string variables are explained near the end of Section 1.6.

A command or a sequence of commands may be expressed as a string variable. The commands in the string form can be run by **eval**. For example,

```
>> s = 'a=5.33;b=2.2;y=sin(a)*exp(-b)'
>> eval(s)
```

or more directly

```
>>eval('a=5.33;b=2.2;y=sin(a)*exp(-b)')
```

or

```
>>eval('a=5.33;'); eval('b=2.2; ');
>>eval('y=sin(a)*exp(-b)')
```

yields the results of calculating y=sin(a)exp(-b) with the defined values of a and b:

$$y = -0.090334$$

This feature may not seem so useful at a glance, but it becomes a powerful means if the commands are to be automatically developed within a program. Fore more information, read Appendices 1 and 2.

Complex variables

A complex variable may be defined by writing i or j next to a number, for example

```
>>a = 5 + 2i
```

The response is

a = 5 + 2i

Of course you can write as >>a = 5 + 2*i.

For the arithmetic calculation of

>>f = 1 + 2i; g = 1 - 2i;
>>h = f*g

we get

h = 5

However, writing as

>>a = b + ci

does not work, where b and c are variable names. You have to write as follows:

>>a = b + c*i

or

>>a = b + i*c

How to find what variables have been used
To find a list of user-defined variable names, use the command

>>who

then a sample response would be

Variables in the current scope:
a   an   f   g   h   s   t

More detailed information may be obtained by **whos**:

>>whos

which is responded as

Variables in the current scope:

Attr Name		Size	Bytes	Class
==== ====		====	======	======
c	a	1x1	16	double
	an	1x12	12	char
c	f	1x1	16	double
c	g	1x1	16	double
	h	1x1	8	double
	s	1x36	36	char
	t	1x1	8	double

Total is 53 elements using 112 bytes

1.5 Looped computation with for/end

Octave provides **for/end** and **while/end** loops. In this section only the former is explained. For quick illustration, the following script calculates $y = x^2 - 5x - 3$ for each of $x = 1, 2, .. 4$ in increasing order:

```
>> for x=1:4
      y=x^2 - 5*x - 3
   end
```

Notice that **for** is terminated by **end**. In the first cycle, x is set to 1, and y is calculated. In the second cycle, x is set to 2 (with an increment of 1), and y is calculated. The same is repeated until the calculation of y is completed with the last value of x. In writing the above statements on command line, Octave does not issue any new prompt sign >> until end is written and *Return* is hit.

The response of Octave to the foregoing statements is

y = -7

y = -9
y = -9
y = -7

If **x** is to be changed with a different increment, the increment can be specified between the initial and last number as follows:

```
>> for x=1:0.5:4
        y=x^2 - 5*x - 3
    end
```

where the increment is now 0.5. The response is

y = -7
y = -8.2500
y = -9
y = -9.2500
y = -9
y = -8.2500
y = -7

Order of calculations in the loop can be reversed as follows:

```
>> for x=4:-1:1
        y = x^2 - 5*x - 3
    end
```

Here, the middle number -1 in 4:-1:1 is a (negative) increment (or decrement) in changing x.

The sequence of x does not have to be constantly incremented or decremented. Any sequence of x values may be used as follows. Suppose the computation is desired to be in the written order of x=2, 0, 15, and 6. The **for/end** loop may be written as

```
>> for x=[2, 0, 15, 6]
        y = x^2 - 5*x - 3
    end
```

where x = [2,0,15,6] looks like an array variable introduced in the next section, but because it appears in a **for/end** loop, it simply means x is sequentially set to 2, 0, 15, and 6 in the order written. The result of the calculation is

$$y = -9$$
$$y = -3$$
$$y = \quad 147$$
$$y = \quad 3$$

1.6   Array variables

One-dimensional array variables

One-dimensional array variables are in a column or a row form, and are closely related to vectors and matrices. In linear algebraic computations, row array is used as *row vector,* and column array is used as *column vector.* The variable x can be defined as a row array by specifying its elements, for example, by

>> x = [0, 0.1, 0.2, 0.3, 0.4, 0.5]

which is responded as

x =
  0.00000   0.10000   0.20000   0.30000   0.40000   0.50000

To print a particular element, type x with its index or position. For example, typing >>x(3) yields

ans =   0.20000

An equivalent way of defining the same x is

```
>> for k=1:6
      x(k)=(k-1)*0.1;
   end
```

22

or

```
>> for k=0:5
       x(k+1)=k*0.1;
   end
```

The size of array does not have to be pre-declared as it is adjusted automatically. The number of elements of x can be increased by defining additional elements, for example,

```
>> x(7) = 0.6;
```

A row array variable with a fixed increment or decrement may be defined as

```
>> x = 2:-0.4:0
```

It yields

$$x = 2.0000 \quad 1.6000 \quad 1.2000 \quad 0.8000 \quad 0.4000 \quad 0.0000$$

The definition of a column array is similar to a row array except that the elements are separated by semicolons; for example,

```
>> z = [0; 0.1; 0.2; 0.3; 0.4; 0.5];
```

An alternative way of defining the same thing is to put a prime after a row array:

```
>> z = [0, 0.1, 0.2, 0.3, 0.4, 0.5]'
```

The prime operator is the same as the transpose operator in the linear algebra, so it converts a column vector to a row vector and vice versa. Typing z as a command yields

```
z =
    0
    0.1
```

0.2
0.3
0.4
0.5

If a single element of an array **c** is specified, for example,

>> c(8) = 11;

c(k)=0 is assumed for k=1 through 7, so typing **c** yields

c =
0   0   0   0   0   0   0   11

A new command >>**c =2.2** changes the definition of **c** and responded as

c = 2.2

which is a scalar variable now. If **c** is again redefined as

>> c(1:2:7) =5

it is responded as

c=
5 0 5 0 5 0 5

Array variables can be combined, for example:

>>x=[1 2 3]; y=[0.5 0.6 0.7]; z=[x y]

yields

z = 1   2   3   0.5   0.6   0.7

Some entries may be eliminated from an array, for example:

$$>>z = [z(1:2), z(5:6)]$$

deletes the 3rd to 4th entries from z:

$$z = 1 \quad 2 \quad 0.6 \quad 0.7$$

Two-dimensional array variables

Stacking multiple row arrays of the same size, a two-dimensional array can be created. Likewise, a two-dimensional array may be created by a row array of column arrays. For example, a=[1, 2, 3, 4] and b=[9, 8, 7, 0] are two row arrays of the same length. A new column array may be written as c=[a; b], which becomes a two-dimensional array. To demonstrate creation of the two-dimensional array on the computer, we set

$$>> a=[1, 2, 3, 4], \quad b=[9, 8, 7, 0]$$

and get

$$a =$$
$$1 \quad 2 \quad 3 \quad 4$$

$$b =$$
$$9 \quad 8 \quad 7 \quad 0$$

Now >> c=[a; b] yields

$$c =$$
$$1 \quad 2 \quad 3 \quad 4$$
$$9 \quad 8 \quad 7 \quad 0$$

Here c is a two-dimensional array (or 2-by4 array). Of course, c can be created directly by >> c = [1, 2, 3, 4; 9, 8, 7, 0].

Combining multiple two-dimensional arrays into one two-dimensional array works if all the arrays have the same height. To show an example, we define another two-dimensional array as

>>d=[0.1 0.2; 0.5 0.6]

Then it can be combined with c that is already defined:

```
>>c=[d,c]
c =
    0.10000    0.20000    1.00000    2.00000    3.00000    4.00000
    0.50000    0.60000    9.00000    8.00000    7.00000    0.00000
```

Some columns of c can be deleted, for example:

```
>>c=[c(:,1:2), c(:,6)]
c =
    0.10000    0.20000    4.00000
    0.50000    0.60000    0.00000
```

Looped calculations using array variables

With array variables, repeating the same calculations for different input becomes easy and efficient. For example, suppose we wish to compute the values of $y=x^3 + 2x^2 - x$ for x=2.2, 3.1 and 5.0. A script to do the entire calculations with a **for/end** loop is

```
>> x=[2.2, 3.1, 5.0];
>> for k=1:3
        y(k) = x(k)^3 + 2*x(k)^2 – x(k);
   end
```

Addition and subtraction of array variables

Array variables can be added and subtracted as long as all array variables are in the same length and same shape. Some examples are

```
>> x=[1 3 0 1]; y=[1 2 2 1];
>> z=x+y
z=  2    5    2    2

>> z=x-y
z=  0    1   -2    0
```

```
>> x=[1 3 0 1]'; y=[1 2 2 1]';
>> z=x+y
z=   2
     5
     2
     2

>> x=[1 3; 0 1]; y=[1 2; 2 1];
>> z=x+y
z=   2   5
     2   2

>> z=x-y
z=   0   1
    -2   0
```

Mathematic functions with an array argument

If an array variable is used as the argument of a mathematical function, the result becomes an array. For example,

```
>>x=[1 1.2; 0.5, pi/4];
>>y=tan(x)
```

yields a 2-by-2 array of tangent values as

```
y =
    1.55741    2.57215
    0.54630    1.00000
```

Here, tan(x) itself is an array the size of which is identical to that of x. This becomes important when the function is used in array arithmetic multiplication and division.

Array arithmetic multiplication and division

We first consider a sample computation with **for/end** statements:

```
>>x=[2.2,3.1,5.0]; y=[1,3,5]; z=[1,1,2];
```

```
>>for k=1:3
        u(k) = x(k)*y(k)+ 2*x(k)^2 –x(k)/z(k);
      end
```

With array arithmetic operators, the above script is equivalently written more compactly as

```
>>x=[2.2,3.1,5.0]; y=[1,3,5]; z=[1,1,2];
>>u = x.*y + 2*x.^2 – x./z;
```

where **.***, **./** and **.^** are called respectively array multiplication, array division and array power operators.

Another example of array arithmetic division is as follows:

```
>>x=[1,3; 4 5]; y=[2,3; 4,2];
>>z=x./y
```

yields

```
z =
    0.50000    1.00000
    1.00000    2.50000
```

Here, the division operation takes place to the members of the array variables in the same position, and the computation is equivalent to

```
for i=1:2
  for j=1:2
     z(i,j)=x(i,j)/y(i,j);
  end
end
```

A function with an array variable as an argument becomes an array of the function. In the following example, $\tan(y)$ and $\exp(x)$ both are 2-dimensional arrays, for example:

```
>> x=[1 3; 0 1]; y=[1 2; 2 1];
>> z= tan(y)./ exp(x)
```

yields

```
z =
     0.57294   -0.10879
    -2.18504    0.57294
```

The above is equivalent to

```
>> for m=1:2
       for n=1:2
           z(m,n)=tan(y(m,n))/ exp(x(m,n));
       end
   end
```

Array arithmetic power operator

The arithmetic power operator, ".^", applies to each member of an array variable, for example:

```
>> x=[2.2, 3.1, 5.0];
>> y = x.^3 + 2*x.^2 – x
```

yields

```
y = 22.528      52.111     180.000
```

and equivalent to

```
>> x=[2.2, 3.1, 5.0];
>>for m=1:3
   y(m)= x(m)^3 + 2* x(m)^2 – x(m);
   end
>>y
```

Array length and array size

29

Array length can be found by the **length** function: for example

>> x=[1, 3, 5, 0, 1, 1];
>> n=length(x)

yields

n = 6

Therefore, if a calculation is repeated for all members of an array variable x, a script may be written with **length(x)** as,

>> for k=1:length(x)
        y(k)=x(k) + x(k)^3;
    end

Use of **length** is limited to one-dimensional arrays, but **size** can be used instead for two-dimensional arrays. Let us see how **size** works with the two-dimensional array of c:

>> c =[1, 2, 3, 4; 9, 8, 7, 0];   size(c)
ans =
        2   4

which means the vertical height of c is 2 and horizontal length is 4. **size(c)** is an array, so >>**f=size(c)** yields **f(1)=2** and **f(2)=4**.

String variables revisited
String variables, already explained earlier, are arrays of characters. The string arrays can be combined just like arrays of numbers.

Here is an example:

>> a='string1', b=' ', c='string2'
>> d = [a, b, c]

yields
a = string1

30

b =
c = string2
d = string1 string2

where b is a string variable of one blank space. See also

>> a(2:4)
ans = tri

Here, a(2:4) is second though fourth characters of a.

Array variables as vectors and matrices

When array variables are used in linear algebra, row and column array variables are called respectively *row* and *column vectors*, and 2-dimensional array variables are called *matrices*. Rules of addition and subtraction of vectors and matrices are the same as mentioned for array variables. Vectors and matrices can be added or subtracted just like ordinary array variables as long as all the vectors or matrices involved are in the same size and same shape.

However, multiplication and division of vectors and matrices are different from those of array variables. The fundamental key is to understand multiplication of vectors as follows.

If u is a row vector and v is a column vector of the same size, the product, [row vector]-by-[column vector] becomes a scalar. For example,

>> u=[1 2 3]; v=[4 5 6]';
>>w=u*v

yields

w = 1x4 + 2x5 + 3x6 = 32

The above computation is equivalent to

>>w=0;

```
>>for m=1:3
    w=w+u(m)*v(m);
        end
>>w
```

If u is a stack of row vectors, the foregoing multiplication changes to:

```
>> u=[1 2 3; -3 -2 -1]; v=[ 4 5 6]';
>>w=u*v
```

yields
```
    w =
        32
        -28
```

Here the multiplication of row vector times column vector took place twice, first for the first row of u times v, and second for the second row of u times v.

What happens if u is a row vector, but v becomes a matrix with two columns? See the following example:

```
>> u=[1 2 3]; v=[ 4 5 6; 3 1 2]';
>>w=u*v
```

```
    w =

        32      11
```

Here, row times column took place twice, first for u times the first column of v, and then for u times the second column of v, that yields a row vector w.

So, if u is a 2x3 matrix and v is a 3x2 matrix, how the product would go? The answer is the combination of two foregoing examples plus the second row of u times the second column of v, as follows:

w =
      32   11
     -28   -13

We write here two more examples with two 3x3 matrices and 3 column vector:

```
>> u=[1 2 3; -3 2 0; 2 5 1]; v=[ 2 1 0; 1 2 1; 0 1 2]';
>> w=u*v
w =
      4      8      8
     -4      1      2
      9     13      7

>> u=[1 2 3; -3 2 0; 2 5 1]; v=[ 2 1 0]';
>> w=u*v
w =
      4
     -4
      9
```

Finally about division by a matrix in linear algebra. There is no straightforward division of a matrix by another matrix but the closest thing is multiplication by the inverse of a matrix. Inverse of a square matrix c is written as c^{-1}, which is calculated by inv(c), and has the following properties:

$$c\,c^{-1} = c^{-1}\,c = I$$

where $cc^{-1} =$ is multiplication of c-by-c^{-1}, likewise $c^{-1}c$ is c^{-1}-by-c, and I is the identity matrix in which all the entries are zero except its diagonal elements are all unity.

Considering multiplication of two matrices, w = uv where u and v are both matrices of the same size, the inverse of this operation is either pre-multiplication by u^{-1} or post-multiplication by v^{-1}:

Pre-multiplication $\quad u^{-1} w = u^{-1} uv \quad => \quad v = u^{-1} w$
Post-multiplication $\quad w v^{-1} \quad = uv\, v^{-1} \quad => \quad u = w v^{-1}$

The readers should try these multiplications using u and v defined earlier.

The foregoing discussions leads us to the solution of a linear equation,

$Ax=y$

where A is a n-by-n matrix, x is unknown n-vector, and y is a known n-vector. The solution of the equation is

$x=A^{-1}y$

However, another equivalent way of writing the solution is recommended:

$>>x=A\backslash y$

For example,

$>>A=[2\ 1;\ 2\ 3];\quad y=[5\ 1]\ ';\quad x=A\backslash y$

yields
        x =
            3.5000
            -2.0000

1.7 Programming with Octave

Being able to develop programs (or scripts) as m-files is one of the most powerful aspects of Octave. Any computational work may be done on the command window, but if the steps are very long or the same computational job is repeated over and over, working only on the command window is tedious and not only inefficient but, in case mistake happens, correction is often impossible.

With a program developed for such lengthy computations, mistakes in the programming can be corrected easily on a program editor and the program (or script) can be run repeatedly any number of times desired. Some effort to learn (1) how to write a program and (2) how to manage the program in the filing system in the computer is necessary, but its benefit overweighs the pain of learning.

An Octave program is a sequence of commands written in a script and saved in a directory as an m-file, which is saved as a file with extension **.m**, or **file_name.m**.

When to program, utilizing GUI (graphic user interface) illustrated in Figure 1.3 is recommended, although other way of writing without GUI and running the programs from CLI is possible. A GUI can be opened by clicking on the GUI icon that appears in the desktop screen of a PC.

If GUI icon is not found in the desktop, click on the *Start* icon at the left bottom corner of the screen. Then, a small window opens, in which click on *All Programs* to open a list of all the programs in the computer. Find the folder named Octave 4.0.0, and click on Octave (GUI).

The GUI has two major columns. A small box at near top of the left column shows the current directory, which in Figure 1.3 is **c:/users/owner**. (The current directory can be **c:/users/owner/ MyOctavePlace** if you have made it the Home directory of your Octave works. For the remainder of this subsection, we assume that this change has not been done.)

The directory can be changed as follows: To go to the parent directory, click on the arrow sign ↑ next to the box showing the current directory. The box under the current directory name lists the names of files or sub-directories in the current directory. So by clicking on one of the subdirectories, the clicked subdirectory opens and becomes the current directory.

Before writing any program, it is recommended to create a new working directory, where you work and save any program you write. The working directory would be in c:/users/owner, but if you wish you can choose any appropriate and convenient place. How to create and open a new directory is explained next.

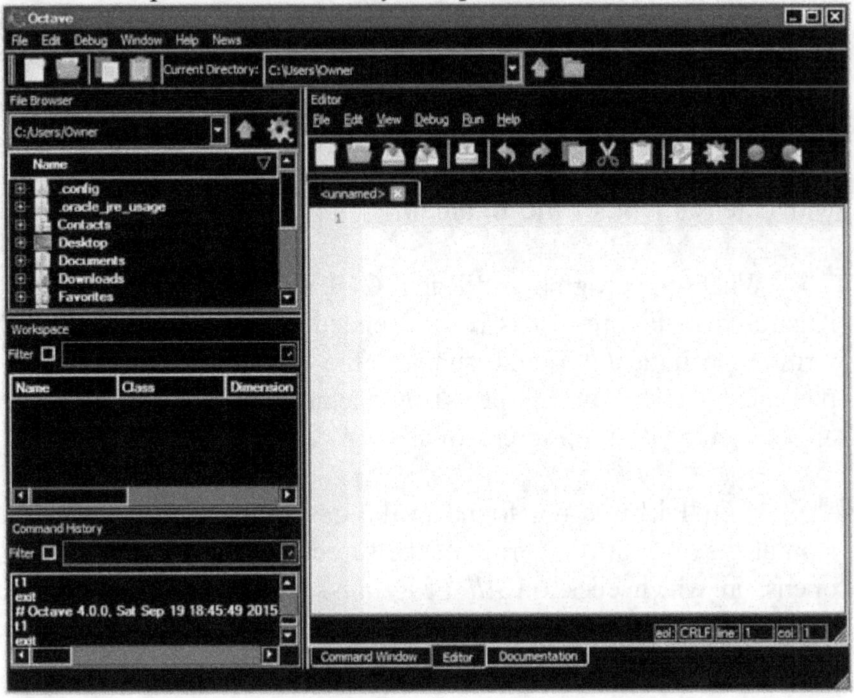

Figure 1.3   Illustration of GUI

Suppose we create a new directory in the current directory, **c:/users/owner**. By clicking on the gear sign next to the thick green arrow sign, a drawdown menu opens, in which an icon for *New Directory* can be found. Click on that, and a small box opens for writing the name of the new name. After writing the name of the new directory, click OK. Then, the new directory is created and appears in the box of files and subdirectories. Click it to open, and it becomes the current directory. If you have not created the directory **MyOctavePlace**, you may do so now using the procedure just mentioned here.

The right column of GUI contains a large box, which can be *Command Window, Editor,* or *Documentation List.* To see which one is currently on the screen, look at the labels under the large area. In the illustration of the GUI in Figure 1.3, Editor is open. To open another item, just click on the selected label.

On the Editor opened, a program can be written with any number of lines. The name of file opened in the Editor is shown at the top left of the Editor. If the name there is unnamed, you will have to give the name before your program is saved and run.

Write a very short program consisting of one or two lines, for example:

```
surf(peaks)
disp('Look at the graphic window to see results.')
```

This script can be run by clicking *Run* in the menu above the editor window. There are two choices in the drawdown menu of Run. The first is *Save File and Run*, and the second is *Run Selection*. If you select the former, you must give the name first, say **prg1**, in the space that opens up. The program created is in the form of **prg1.m** with the suffix **m**. After this, the program starts running if it has no typo or errors. You have to open the Command Window of GUI to read the results because the results can be found only in the Command Window illustrated in Figure 1.4.

If *Run Select* is clicked, a selected part of the program is run without saving the program, where selection means you highlight the part by mouse. This is often a convenient way of testing a part of the program.

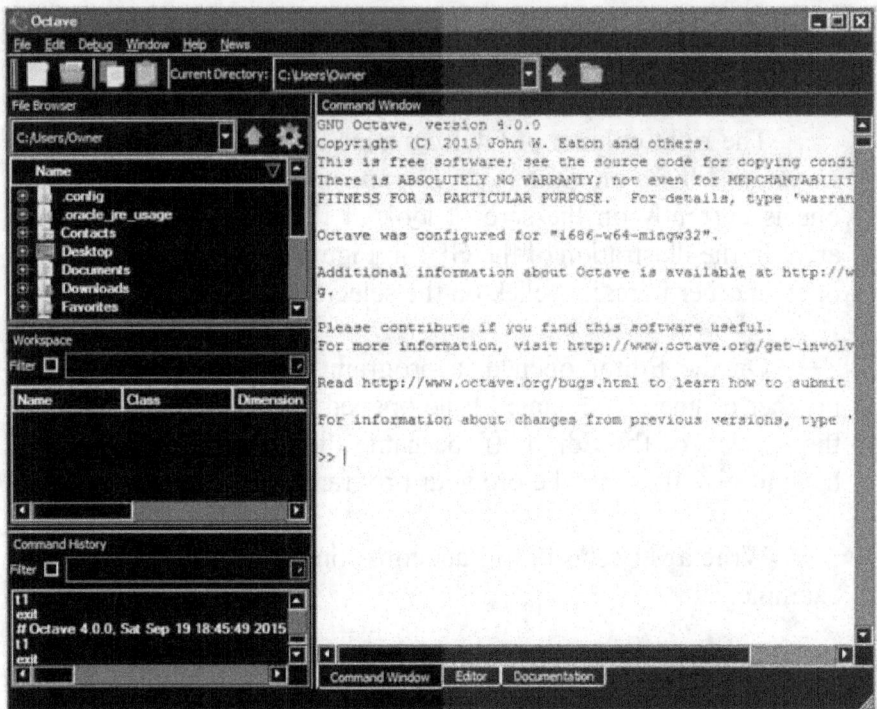

Figure 1.4   Illustration of Command Window of GUI

Caution

The two-line script suggested in the prior subsection will run for sure, but *Run* for your own programs may not succeed in early attempts because of errors. For the errors, relatively detailed messages appear on the Command Window. Sometimes, however, only a weird message may appear. In the worst cases the whole system of Octave may stop and freeze. If this happens, don't hesitate to close Octave and start again. If Octave does not close, you may have to shut down the PC and restart.

Such a nasty behavior most likely starts when the program has some errors. Therefore, it is advised to write only a short part of the program at a time and run that part by *Run Selection*. When that part runs smoothly, add another small number of lines. Only after the program is thoroughly tested part by part, run the entire program by *Save File and Run*. (See Appendix 3 for more about possible pathological behavior.)

Another way of creating a program in the form of xxx.m

The m-file form of a program can be created without using Octave as follows. Once created, it can be opened in the GUI Editor by a simple clicking.

The PC Windows system has a text editing software named *Notepad*, which can be found and opened like any other programs on the PC by clicking on *All Programs* (click first on the *Start* icon on the bottom left corner of PC). *Notepad* may be in *Accessories* in *All Programs*. When a program is written in a different editor such as *Notepad*, test the program in piece meal as mentioned in Caution subsection. One kind of possible trouble with a program written outside the GUI Editor is that some characters or symbols, such as single quote sign, may not be accepted by Octave.

Write your program on *Notepad*, and save it, using *Save As*, with an appropriate name with suffix **m**, like **t3.m**, in the folder of your choice.

Now open the folder where the program, **t3.m**, just created is saved. The m-file created has an Octave icon in front:

Clicking on it will open the m-file in the Octave GUI Editor.

Echo on, echo off

When an m-file is executed, the statements in the m-file are not usually printed on the screen. After **echo** is turned on with the **echo on** command, however, the statements are printed. By doing this, the user can see which part of the m-file is being executed. To turn off echo, type **echo off**. This feature may become useful when debugging an m-file.

Comment statements

39

The percent sign **%** in m-files indicates that any statements after this sign on a line are comments and are ignored in the computations.

1.8 Branch statements with if, else, elsif

The **if** operator allows switching between two choices or more of actions in the computational flow. An example is shown next:

```
if n<= 5, price=15;
    else price=12;
end
```

Here we assume that the value of n was defined before. Then, if n is less than or equal to 5, **price** is set to **15**, and else (otherwise) **price** is set to **12**. The **if** statement is always closed with **end** operator.

The following example is a little more elaborate:

```
if n<= 5, price=15;
    elseif (n>5 & n<10), price=12;
    else price=10;
end
```

Here the second line of **elsif** has been added. The sign **&** means "and". Therefore, if **n** is less than or equal to **5**, **price** is set to **15;** else if **n** is greater than **5** and less than **10**, **price** is **12**; otherwise (that is, if **n** is equal to or greater than **10**) the price is **10**.

Table 1.3 shows all the operators that can be used in **if** statements.

Table 1.3   Operators associated with **if** operator

Operators	Meaning

>	Greater than
<	Less than
>=	Greater than or equal to
<=	Less than or equal to
~=	Not equal to
==	Equal to
|| or |	Or
&& or &	And

NOTE: The && and || operators are not accepted in Matlab, where the correspondent operators are only & and |.

The *not equal* operator is written as "~=":

```
if r ~= 3, vol = (4/3)*pi*r^3;
end
```

Example of *or* operator:

```
if g>3 | g<0, a = 6;
end
```

Example of *and* operator:

```
if a>3 & c<0, b=19;
end
```

Example of *equal* operator: Assuming **r** and **d** are integers defined earlier. The following statement finds out if **r** is divisible by **d**, or not. Namely if divisible, r is printed out:

```
if fix(r/d) == r/d, r
end
```

or equivalently

```
if mod(r,d) == 0, r
end
```

(See Table 1.1 for the **mod** function.)

The **&** and | operators can be used in a clustered form, for example,

```
if ((a==2 |   b==3) & c<5) g=1;
end
```

The **if/end** statement can be inserted in a **for/end** loop or another **if/end** loop. In the following example, **y=sin(x)** takes effect but, if **sin(x)** becomes negative, y is set to 0:

```
for x=0:0.1:3
   y=sin(x);
   if y<0, y=0; end
   [x,y]
end
```

Here a bit of new style of printing x and y is added, that is **[x, y]**. In this way both **x** and **y** are printed in a single row on the screen.

Printing in the foregoing example may be improved by using **disp** command as follows:

```
for x=0:0.1:3
   y=sin(x);
   if y<0, y=0; end
   disp([x,y])
end
```

In the following example, **c=0** initializes the counter **c** to zero, **x =...** defines an array of numbers, and **length(x)** is the length of the array **x**. In the **for/end** loop the counter **c** is incremented by one if **x(k)** is negative. Finally, **c** becomes the total count of the negative elements.

```
c=0;
x=[-8, 0, 2, 5, 7, 2, 0, 0, 4, 6, 6, 9];
for k=1:length(x)
    if x(k)<0, c=c+1; end
end
c
```

As another example, we write a program that removes the numbers divisible by 4 from an array x. Assume the array x is initially given by

```
x=[-8, 0, 2, 5, 7, 0, 4, 6, 6, 9, 16];
```

The program for this chore is

```
k=0; clear   y
for n=1:length(x)
    if x(n)/4 ~= fix(x(n)/4),
        k=k+1;
        if k==1, y(1)=x(n);
            else   y=[y,x(n)];
        end
    end
end
x=y
```

Here, **y=[y,x(n)]** appends **x(n)** to **y**. The result of the run is

$$2 \quad 5 \quad 7 \quad 6 \quad 6 \quad 9$$

==

Remarks: if/end may be replaced by **if/endif**. Likewise **while/end** by **while/endwhile**, and **for/end** by **for/endfor**. Using **endif**, **endwhile** and **endfor** makes the program easier to read, because with **end** only, the program may become congested with many **end**s. The only problem is that Matlab does not accept **endif**, **endwhile** nor **endfor**. Therefore, if compatibility with Matlab is important, these changes should be avoided.

To reduce confusion in case only **end** is used, it is recommended to indent, by 2 blank spaces, the lines between each pair of **for/end**, **if/end**, and **while/end**, for example:

```
for
  while
    if
      for
      end
    end
  end
end
```

Another approach to increase readability, which is acceptable to Matlab, is to place % after **end** but before **if**, **while** or **for** as illustrated here:

```
for
  while
    if
      for
      end %for
    end %if
  end %while
end %for
```

===

Break

The **break** terminates the execution of a **for** or **while** loop. When used in nested loops, only the immediate loop where **break** is placed is terminated. In the next example, **break** terminates the inner loop as soon as $n>2*m$ is satisfied once, but the loop for m is continued until the loop of $m=1:6$ is terminated:

```
for m=1:6
```

```
    for n=1:20
        if n>2*m, break,
        end
    end
end
```

In a programming language that has no break command, **goto** is used to break a loop. Octave, on the other hand, has no **goto** command.

1.9   Loops with `while/end`

The **while/end** loops are similar to **for/end** loops. Although the **for/end** loop has a fixed number of loops (repeats), the number of **while/end** loops is controlled by a condition specified next to **while**. A prior example of finding the number of negative elements in **c** is rewritten with **while/end** now:

```
c=0;
x=[-8, 0, 2, 5, 7, 2, 0, 0, 4, 6, 6, 9];
k=1;
while k<=length(x)
    if x(k)<0, c=c+1;
    end
    k=k+1;
end
c
```

Sometimes a loop that can continue infinitely is used, which may be terminated when a certain condition if any is met. The following example is an infinite loop that is terminated only if the condition **x>xlimit** is met:

```
while 1
    --- some calculation of x here ---
    if x > xlimit, break; end
end
```

Here **1** after **while** means that **while** is to be continued without any condition, so the loop continues on and on until the break statement is satisfied by a chance. Don't run any infinite loop unless some mechanism to stop the loop is built in. If an infinite loop continues, the only way to stop is to shut down the software or computer, provided that you know how.

1.12 Output technique

Formatted output

In all of the foregoing examples, the results of computations are printed in a very primitive way, that is, just numbers. However, the style of writing can be refined as follows using **fprintf**.

In the following example, the x and y values are printed using the **fprintf**:

```
for x=1:4
   y=x^2 - 5*x – 3;
   fprintf('x = %f   y = %f\n', x, y )
end %for
```

where **%f** specifies the format to be the f-format (floating point value format), and **\n** is the line feeder. The result is

```
x = 1.000000   y = 1.000000
x = 2.000000   y = 4.000000
x = 3.000000   y = 9.000000
x = 4.000000   y = 16.000000
```

To see the effect of **\n**, delete it and run to see what happens. You will see the printout is jammed without line feed.

To use the e-format, change **%f** to **%e**:

```
for x=1:4
```

```
        y=x^2 - 5*x – 3;
          fprintf('x = %e   y = %e\n', x, y )
        end %for
```

The result is

```
        x = 1.000000e+000   y = 1.000000e+000
        x = 2.000000e+000   y = 4.000000e+000
        x = 3.000000e+000   y = 9.000000e+000
        x = 4.000000e+000   y = 1.600000e+001
```

When to print an integer, the %i format may be used, for example

```
        z=50
        >> fprintf('%i \n', z)
        50
```

However, when z is a floating point value, the format is automatically changed to %f:

```
        z=pi*100
        >> fprintf('%i \n', z)
        314.159
```

In using the %f and %e, the number of decimal places may be specified by changing them to, for example, %.3f and %.2e respectively:

```
        >> fprintf('%.1f   %.5f \n ', z, pi)
        314.2   3.14159
```

```
        >> fprintf('%.1e   %.5e \n ', z, pi)
        3.1e+002   3.14159e+000
```

disp

Command **disp** displays a number, an array variable, or a string on the command window without variable name. Therefore, it may be used to neatly display messages or data flexibly on the screen. For example, **disp([pi, 2.2, 4.1])** prints

3.14159   2.2000   4.100

on the command screen.

The **disp** command is useful to warn the user of a program before **input** command is used by instructing how to write the input, for example:

disp('The following input must be a string enclosed by single quote signs')

The **disp** command even evaluates an equation, for example

>>x=pi/4; disp(y=sin(x)*cos(x))

calculates the equation, **y=sin(x)*cos(x)**, and sets the value of **y**. If the equation is written directly like >> **y=sin(x)*cos(x)**, "**y=**" is printed before the value of y, but **disp(y=sin(x)*cos(x))** simply displays the value of **y** without "y=". Consider using **disp** as:

x=pi/4;
disp('sin(x)*cos(x)=')
disp(y=sin(x)*cos(x))

The response is

sin(x)*cos(x)=
0.50000

An example of displaying the result of multiple equations is as follows:

x=pi/4; disp([y1=sin(x), y2=cos(x), y3=tan(x)])
0.70711   0.70711   1.00000

sprintf

It is very similar to **fprintf** except that **sprintf** writes the output into a string.

This statement is often used to create a command in a string that can be executed as **feval(s, a)**, where **s** is a string of a function name, and **a** is the argument of the function. For example:

```
>>s=sprintf('tan');   feval(s,1)
ans =   1.5574
```

It is useful when a command is to be created or edited automatically and executed within an m-file.

Writing into a file

With **fprintf**, it is possible to write into a file rather than printing on the screen. To do this, a named file must be opened by **fopen**. The following example illustrates the procedure:

```
vol=55.8;
file_id = fopen('file_x', 'w')
fprintf( file_id, 'volume= %f\n', vol)
fclose(file_id)
```

Here, **file_x** is first opened, and the value of **vol** is written in the file using the f-format. Then the file is closed.

1.13   Input technique

To input a value to the program being run is possible by input. A simple example is

```
x = input('Type input for x and hit Return:   ')
```

The result is

```
Type input for x and hit Return:   3.14
```

x = 3.1400

"Type input for x and hit Return: " is the printout on the screen, and 3.14 is what the user typed on the key board. The next x = 3.14 is printed by the computer because the statement of "x = input(...)" is not terminated by the **;** operator.

The input statement can be used anywhere in the script including **for/end%for** and **while/end%while** loops. Here is another example:

```
r=0:
while r<10
    r=input('Type radius(or -1 to stop):');
    if r< 0, break, end %if
    vol = (4/3)*pi*r^3;
    fprintf('Volume = %f\n', vol)
end %while
```

In the foregoing loop, the value of radius **r** is typed on the keyboard. The **fprintf** statement is to print out **vol** with the f-format, **%f**. If **0<r<10**, **vol** is computed and printed out, but if **r<0** the loop is terminated. Also, **if r<10** is dissatisfied once, the while loop stops.

1.14   Writing and reading by save and load

save and load
All the data of the user-defined variables currently in the memory space can be saved by

>> save file_name

which saves all the data in the memory into the file named file_name with or without an extension. The file created can be found in the current directory. Before or just after you save the data, you may want to see what variables are saved. In this case, use >>**who**, which lists all the variable names that are saved into **file_name**.

The file may be opened in the GUI Editor if you need to see the contents. If you use a name like file_name.txt with the extension .txt, the file may be opened in *Notepad*. With extension .doc, the file may be opened in MS Word.

Let us see more details of what happens. For illustration, we create two variables by

```
>> x=[1 2 3 5]
x =
    1    2    3    5

>> y=[9;4;7;1]
y =
    9
    4
    7
    1
```
Then typing

```
>> save sfile
```

creates a file named **sfile**. Here we assume that no other data than x and y are in the memory, so only x and y should be in **sfile**. The contents of this file opened in GUI Editor are:

```
# Created by Octave 4.0.0, Wed Sep 30 08:09:50 2015 Eastern
Daylight Time <unknown@---->
# name: x
# type: matrix
# rows: 1
# columns: 4
  1 2 3 5

# name: y
```

```
# type: matrix
# rows: 4
# columns: 1
 9
 4
 7
 1
```

This file may be loaded any time later by

>>load sfile

Then all the data saved before are recovered in the memory.

Save in ascii and load ascii data file

Data may be saved in ascii format by the **save –ascii** command. For example

>> save –ascii afile x y

saves **x** and **y** in ascii format in the **afile** file..

Assuming that the same **x** and **y** defined in the prior subsection are saved, the **afile** opened on GUI Editor looks like

```
1.00000000e+000 2.00000000e+000  - - -   5.00000000e+000
9.00000000e+000
4.00000000e+000
7.00000000e+000
1.00000000e+000
```

The data saved in ascii format can be loaded by the load command, but loading a file in ascii format is not quite the inverse of **save** in ascii format. The reason is that while **save -ascii** can save multiple variables, **load** reads the entire file as data of only one variable. Furthermore, the file name becomes the variable's name. For example, if a file named **ydat.tmp** is loaded by

>> load ydat.tmp

the content is loaded to the variable named **ydat** regardless of the extension name.

Therefore, the data file **ydat** must contain data in only one of the following data forms:

(a) a single number
(b) a row array
(c) a column array
(d) a two-dimensional array

If multiple variables have to be loaded, each variable should be saved in a separate ascii data file.

Data files prepared by another computing software such as Fortran or C in ascii (or text) format can be loaded by **load filename.extension** as long as the data structure is one of the four forms mentioned above.

Creating file names automatically

A method to create filenames automatically within an m-file becomes desired some times. If a whole command, including the filename, is written as a string, it may be executed by **eval**. In the following script, **xdata** is assumed to be computed for each k and saved in separate **files** named **fname001, fname002, ...** in ascii format.

```
for k=1:kmax
%Here are some statements to produce xdata for each k.
%kmax is the maximum number of k (less than 1000)
If k<10,
      s=['save fname00',num2str(k),'xdata –ascii']
elseif k>=10 & k<100,s=['save fname0',num2str(k)','xdata –ascii']
elseif k>=100,
      s=['save fname',num2str(k),'xdata -ascii']
eval(s)
end
```

1.15 How to write user-defined functions

User-defined functions, which are saved as separate m-files, are equivalent to subroutines and functions in other languages. All the functions, whether it is a built-in function or user-developed function, are written in the form of a function m-file, and they have the same format as follows

function y = function_name(x1, x2, …)

and saved as function_name.m.

Let us consider developing a function to calculate the equation:

$$y = (2x^3 + 7x^2 - 1)/(x^2 + 5e^{-x})$$

Assuming that the name of the function is **my_function**, the whole m-file for this function is

function y = my_function(x)
y = (2*x.^3 +7*x.^2 -1)./(x.^2 + 5*exp(-x));

This function is saved as **my_function.m**. Notice that arithmetic operators are used in case x is an array variable. If x comes as a m-by-n array, the result y is returned as an array of the same size.

If this function is called by

>>z = my_function(3)

the response is

z = 502.1384

If the argument is an array, for example

x = [3, 1; 0, -11]
then,

>>z = my_function(x)

yields

z =
   1.2542e+001   2.8175e+000
  -2.0000e-001   -6.0636e-003

A user-defined function can return multiple values. For example, the following function is written to return two values, **y** and **z**:

function [y,z]=demo_f(x1,x2)
y = (2*x1.^3 +7*x1.^2 -1)./(x1.^2 + 5*exp(-x2));
z=exp(x1).*sin(x2);

Then, an example of using it is

>>[a1, a2]=demo_f(1.1, 4.2)
a1 =   7.8850
a2 =   -2.6184

Function that calls another function

The argument of a function may be the name of another function. For example, suppose a function that evaluates a weighted average of a function for three different values of argument is given by

$$z = (f(a) + 2f(b) + f(c))/4$$

where f(x) is the function still to be named. The following script illustrates a user-defined function **avf.m** that computes the foregoing equation:

function w = avf(fname, a, b, c)
w = (feval(fname,a)+2*feval(fname,b)
      +feval(fname,c))/4;

In the foregoing script, **fname** is the name of the function in a string variable form. For example, if sin(x) function is used for f(x), with a, b and c pre-defined, we write

```
>> s = 'sin'
>> z=avf( s, a, b, c)
```

The z value computed becomes the value of

$$z = (\sin(a) + 2\sin(b) + \sin(c))/4$$

Any function name may be written in place of **sin** once function **avf** is developed.

Debugging of function m-files
Debugging function m-files is more difficult than ordinary script m-files. One reason is that you cannot see the values of variables in function by typing the variable names on the command window. The most basic but effective method of developing a function is to comment out the function statement on the first line by placing **%** before **function** and then test the m-file as an ordinary m-file. Put the function statement back after a thorough examination of the m-file.

Exercise problems for Chapter 1

[1] v=rand(1,100) creates the variable v, which is a row array of 100 random numbers. Make a program to calculate (i) average of the 100 numbers, (ii) determine how many are above 0.5, and (iii) how many are below 0.5.

[2] Calculate the values of $y = x^2 - 2x - 2$ for x= –5, –2, 0, 3, 5 by (i) writing a program using for/end loop, and (ii) not using for/end loop but using the array arithmetic operators.

[3] Write a program to answer Problem [2] such that the function is given by input as a string, and an array of x is given as input.

[4] Write a function that takes x as input, and calculate $y=\exp(x)$ if $x<0$, but $y=1/(1+x^2)$ if $x>0$.

[5] Write a program that generates N random numbers and calculates how many random numbers fall in the intervals of $0 \leq x < 0.2$, $0.2 \leq x < 0.4$, .., $0.8 \leq x \leq 1.0$, and what fraction of N fell into each group. Calculate the variance of the fractions. Run the program for N=1000, N=10000, and N=100000.

[6] Hilbert matrix is an array of a(i,j) defined by

$$a(i,j)=1/(i+j-1)$$

A hilbert matrix is generated by >>**a=hilb(n)**. Write a program that adds 1 to every entry of 5-by5 hilbert matrix and save it in a file named hilb5.txt. After that, clear the memory by **clear all**, and load the file, hilb5.txt, and print it.

[7] Array variables are defined by

u=
1 2 3 4

v=
1
2
3
4

w=
4 5 6 7

a=
1 4 3
2 -1 1

1 0 2

b=
1 0 1
2 1 1
4 1 2

c=
1 0 0 1
2 1 0 1
3 1 1 0
4 1 2 1

Which of the following operations are valid?

u + v
u + w
c*v
c*w
a+b
b*c
a*b
a.*b
u.^2

[8] Two one-dimensional arrays are defined by

a=[1 2 3]
b=[4 5 6]

Some calclulations with a and b become as follows:

>> a*b'
ans =   32

>>a'*b
ans =
        4   5   6
        8   10   12

```
         12     15     18

>> a.*b
ans =
      4     10     18
>> a.^b
ans =
      1     32    729

>> a.*b'
ans =
         4      8     12
         5     10     15
         6     12     18
```

Explain what happened to each.

[9] A row array of number is defined by d=1:29, that is an array of days in February in a leap year. Write a program to make a calendar for February with 7 columns, the first row is Sunday, the second Monday and so on. Assume the first day of February is Monday. The first row may be S, M, T so on, but you may ignore this if it is not easy. The blank days of the calendar may be filled by 0, or blank, which ever you prefer.

[10] An array x has three columns and 10 rows. The first column is numbers starting with 1 and ending at 10, the second column is [0 0.5 1.1 1.6 2.2 2.7 3.4 3.9 4.4 5.0] '. The entries in third column equal those in the second column squared. Print the table as neatly as possible, and print on the top of the table the captions of the columns such as n, x, and x^2.

[11] String variables are defined: A= 'Jones', B= 'Smith', C= 'Alexander', D= 'Xing', E= 'Camden '. Write a program in which all the string variables are first defined. Now, put all of these string variables into one string variable, Z, such that Z is a two-dimensional array of letters, the fist row is Jones, second Smith, and so on.

Chapter 2
Graphics with Octave/Matlab

Graphics plays a central role in Octave. Plotting a given data set or the results of computation is possible with very few commands. Octave allows you to finish scientific as well as business graphics with the highest possible sophistication and elegance.

Trying to understand mathematical equations with graphics is an enjoyable and very efficient way of learning mathematics. Indeed, we might say that *unless you understand mathematical equations graphically, you don't really understand them.* The same applies to scientific and business data. Being able to plot mathematical functions and data freely is the most important step, and this chapter is written to assist you to do that.

For professional people, the importance of graphics is even more profound. Massive data generated today by computers and experiments, as well as from business information source, can only be effectively presented by means of graphic visualization. With Octave, it is achieved by plotting data in appropriate forms including one-, two-, and three-dimensional plots, or some times motion pictures (four-dimensional plots).

The best way to learn graphic commands is to read a small amount of instructions at a time and then practice on Octave, first by typing examples in the command window or executing the script m-files of this chapter and, second, by changing the m-files in various ways.

2.1 How to plot

plot
Suppose a set of data points, (x(k),y(k)), k=1, 2,..., n, is to be plotted, where x(k) is an abscissa value and y(k) is an ordinate value. You need to prepare x and y in an identical array form;

namely, x and y are row arrays (or both column arrays) of the same length. Then, the data may be plotted using **plot(x,y)**.

As an example, we plot a function:

$$y = \sin(x) \exp(0.4x), \qquad 0 < x < 10$$

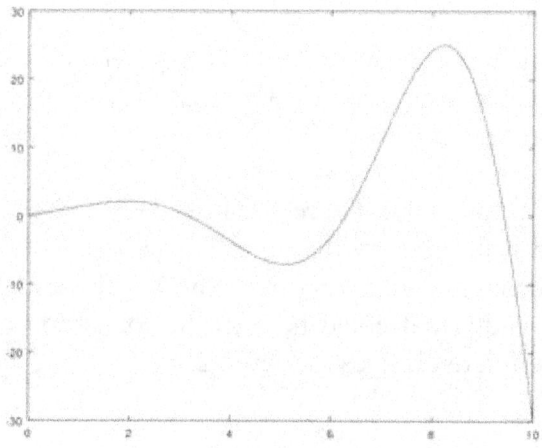

Figure 2.1   A bare bone figure

The following script is a quick attempt:

```
List 2.1
clear, clf, cla
n=201;
delx=10/(n-1);
for k=1:n
   x(k)=(k-1)*delx;
   y(k)=sin(x(k))*exp(0.4*x(k));
end %for
plot(x,y)
```

or more compactly:

```
List 2.1A
clear, clf, cla
x = 0: 0.05: 10;
```

61

```
y=sin(x).*exp(0.4*x);
plot(x,y, 'linewidth',4)
```

Each of the foregoing scripts plots Figure 2.1.

We assume in the foregoing scripts that the number of data points is 201, then the increment of the x, namely **delx**, equals 10/200. Notice in the preceding list, x is a row array of 201 abscissa values of 0, 0.05, 0.1, ..., 10, and y is a row array of ordinate values of the same length. Then, **plot(x,y)** plots the data. The value of n is selected arbitrarily. However, if n is too small, the plotted function loses smoothness.

Figure 2.1 is a bare bone figure because there are no axis labels, no titles, and lines are very thin to see. The line width may be increased by changing **plot(x,y)** to **plot(x,y,'linewidth',4)** where 4 is the line width that could be a smaller or larger value. With this change the figure changes to Figure 2.1A:

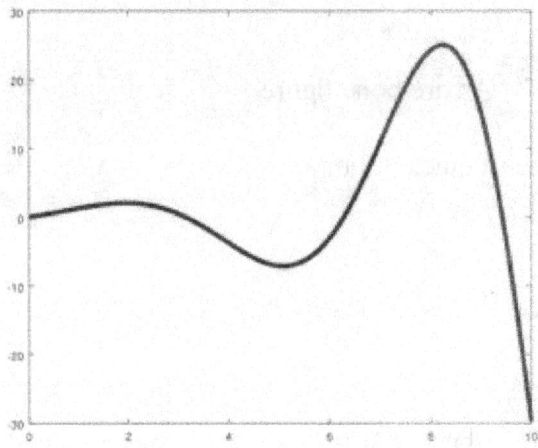

Figure 2.1A   The line width is increased

The default value of the **linewidth** is 0.5. If this is the desired line width, **linewidth** needs not be stated and writing simply **plot(x,y)** suffices.

Further improvements may be done by adding axis labels with a large font size, and making the axis lines thicker. An improved script is List 2.1B, which produces Figure 2.1B.

```
List 2.1B
clear, clf, cla
x = 0: 0.05: 10;
y=sin(x).*exp( 0.4*x);
y=sin(x).*exp( 0.4*x);
plot(x,y, 'linewidth',4)
xlabel('x','fontsize',18)
ylabel('y','fontsize',18)
set(gca,'fontsize',16);
set(gca, 'linewidth',2)
```

The axes are labeled by **xlabel('x',fontsize',18)** and **ylabel('y', 'fontsize',18)**. The last two lines in List 2.1B are to make the coordinate lines and tic values thicker, namely the font size of tick values is set to 16 and coordinate line width to 2. These two **set** statements are not necessary in case the default font size, line width of the coordinate lines and tic marks are acceptable. In this book, all figures are reduced in size at the time of printing, so without the increase in **linewidth** and **fontsize**, the figures would be too thin to read.

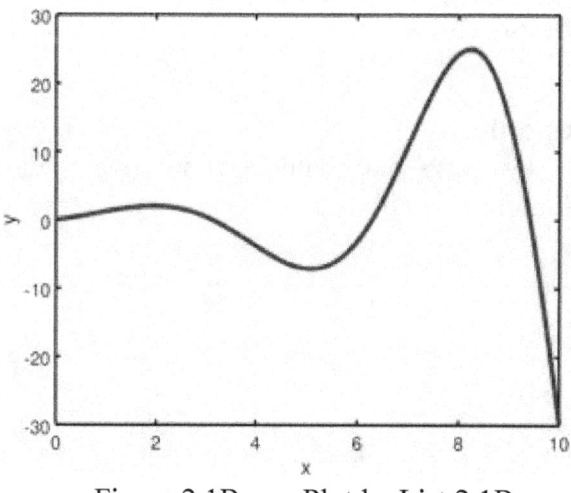

Figure 2.1B   Plot by List 2.1B

63

Figure 2.2 is plotted by List 2.2 connecting a series of points in a complex plane.

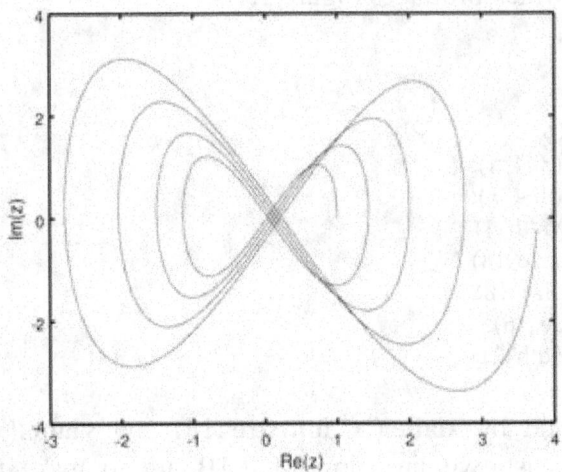

Figure 2.2   Plot on a complex plane using List 2.2

```
List 2.2
clear, clf, cla
p=0:0.05:8*pi;
z=(cos(p) + i*sin(2*p)).*exp(0.05*p) + 0.01*p;
plot(real(z), imag(z))
xlabel('Re(z)','fontsize',16)
ylabel('Im(z)','fontsize',16)
```

Plotting by marks only

Data can be plotted by marks only without connecting by lines. Nine samples of marks are illustrated here:

Point :   .
Plus :   +
Star :   *
Diamond: d
Circle :   o
Pentagon : p
Square :   s
Triangle :   ^

x-mark :   x

To find more marks, type >>**help plot**. To plot with one type of mark only, place the mark symbol as a string after the coordinates in the arguments of **plot**. The graph produced by List 2.3 is shown in Figure 2.3.

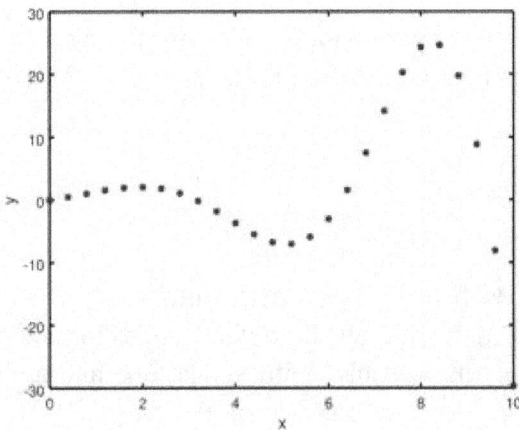

Figure 2.3   A graph plotted with marks only

List 2.3
Clear, clf, cla
x = (0:0.4:10);
y=sin(x).*exp(-0.4*x);
plot(x,y,'*', 'linewidth', 4);
xlabel('x','fontsize',16)
ylabel('y','fontsize',16)
set(gca,'fontsize',14);
set(gca, 'linewidth',2)

Line types
Four line types are available as shown next:

Line Type Symbols

Line type	Symbol
Solid	-
Dash	--

Dotted	:	
Dashdot	-.	

Default is the solid line type. To plot with a selected line type, specify the line mark after x and y, for example:

    plot(x,y,'-.')

If thick lines are desired, add **'linewidth', n** before the closing the right parenthesis, where **n** is a line width number such as 2, 4 or 6, for example,

    plot(x, y, '-.', 'linewidth', 4)

Plotting a function with both lines and marks

Plot twice, namely, the first time by the default solid line and the second time with marks only. To plot with a solid line and a mark +, for example, write the plot command as **plot(x,y,x,y,'+')**, where '+' applies to the second pair of **x** and **y**. The text command may be used to plot with any mark or letter; however, the location of the mark may be slightly offset from the correct location of the data point.

Line colors

Eight colors, namely, red, yellow, magenta, cyan, green, blue, white, and black, are available for the lines and marks. These colors are specified by letters, r, y, m, c, g, b, w, and k, respectively. Use the color symbol just like the line types in the argument of plot, for example:

    plot(x,y,'g')

A combination of mark and color is also possible:

    plot(x,y,'+g')

plots the data with + marks in green.

Cleaning and clearing graph

clf clears everything inside the graphic window, while **cla** clears the plotted curves and redraws the axes. Put **clf** always in the beginning of a program that uses plot in order to prevent confusion with the graph plotted in the previous program.

Figure's position on the computer monitor screen

The command **figure** opens a new graphic window that is numbered consecutively from the previous one, while **figure(n)**, where **n** is an integer, opens the figure window numbered by **n**. If multiple figure windows exist, you have to be aware which one is the current figure. This is because all the graphic commands apply to the current figure. The latest window opened is the current window unless an older one is called for by **figure(n)**. The sequential number **n** is displayed at the top left corner of the figure window.

The size and shape of figure on the computer monitor are determined by default. However, the size, shape and location on the computer monitor may be changed by

figure(n,'position', [pix,piy,pwx,pwy])

where **pix** and **piy** are the horizontal and vertical pixel coordinates, respectively, of the left bottom corner of the figure window in the monitor pixel coordinates (the origin of the pixel coordinates of the monitor is at the left bottom corner of the screen); pwx is the number of pixels in the width; and pwy is the number of pixels in the height of the figure window. By specifying **[pix, piy, pwx, pwy]** appropriately, a desired location on the monitor screen, size and shape may be achieved. The existing figure may also be changed by **set(gcf, 'position', [pix, piy, pwx, pwy])**. The coordinate values of the current figure may be obtained by **get(gcf,'position')**.

With the **figure** command, it is also possible to display multiple figures in neatly organized manner on the monitor.

close

close(n) closes **figure(n)**, and **close all** closes all figures.

axis, axis on, axis off

For a figure, the minimum and maximum of the coordinates, tic marks, and the coordinate values at the tic marks, are all determined automatically. Some properties of a figure may be changed as illustrated next.

The coordinate axes and tic marks can be removed by

axis off

which may be written in an m-files, or typed on a keyboard while the figure is open. The axes and tics are reinstated by **axis on**.

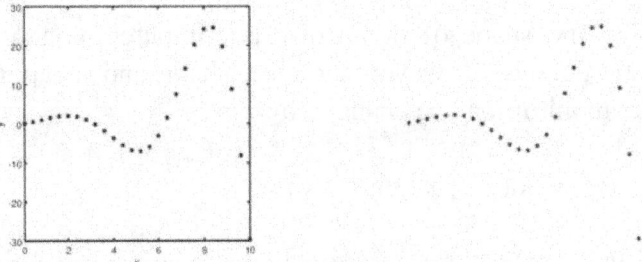

Figure 2.4   A plot with **axis square** (left), and a plot with **axis off** (right)

The maximum and minimum of the coordinates of the graph may be specified by

axis([x-min, x_max, y_min, y_max])

Any lines outside the limits will be clipped. This command can be used in an m-files, but also can be typed on the keyboard so that the view area can be changed as many times as desired while the figure is on the monitor screen. It is suggested that the reader appends **axis([-10, 20, -20, 30])**, as a test trial, to List 2.3 to see the effect of axis.

Grid on, grid off

A grid can be added to the graph by **grid on**, while **grid off** removes the grid. Simply using **grid** multiple times toggles **grid on** and **off**. An example of using **grid on** is illustrated in Figure 2.5 plotted by List 2.4.

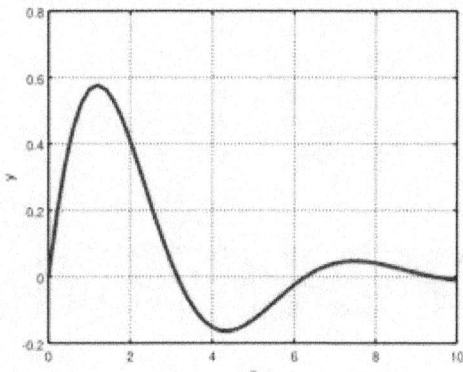

Figure 2.5   A figure with `grid on`

```
List 2.4
clear,clf,cla
x = (0:0.2:10)';
y=sin(x).*exp(-0.4*x);
plot(x,y,'linewidth', 4)
grid on
xlabel('x','fontsize',16); ylabel('y','fontsize',16)
set(gca,'fontsize',14); set(gca,'linewidth',2)
```

Polar plot

A function on a polar coordinate can be plotted by **polar**. Figure 2.6 is plotted by List 2.5.

```
List 2.5
clear,clf,cla
t = 0:.05:pi+.01;
y = sin(3*t).*exp(0.3*t);
```

```
polar(t,y)
title('Polar plot')
grid on
set(gca,'fontsize',14);
set(gca, 'linewidth',2);
```

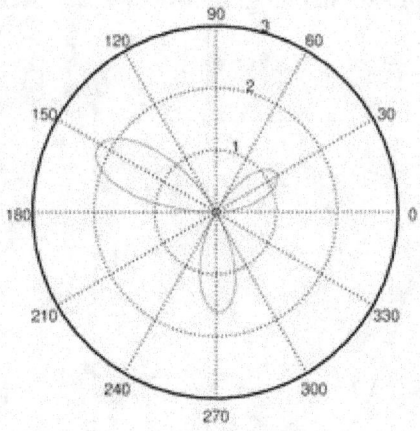

Figure 2.6   Polar plot

Log and semi-log plot

A function may be plotted on a log-log scale by **loglog**. See List 2.6 and Figure 2.7.

```
List 2.6
clear,clf,cla
t = .1:.1:2;
x = exp(t);
y = exp(t.*sinh(t));
loglog(x,y)
grid on
xlabel('x');
ylabel('y')
```

In this m-file, **xlabel** and **ylabel** are used without fontsize, but with the default font size. Also **set(gca ...** is not used. These are intentional for the purpose of illustrating the effect of using the default font size and default line width for the coordinate lines.

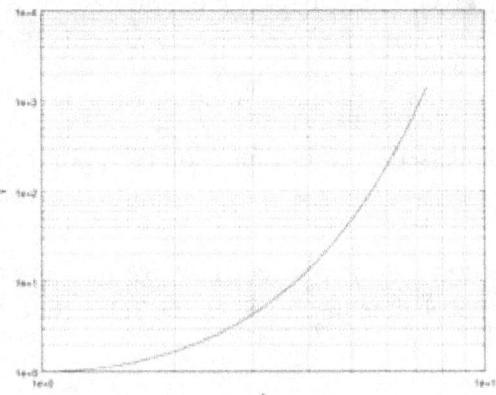

Figure 2.7   A log-log plot

A semi-log plot with the log scale for y is produced by List 2.7. See Figure 2.8.

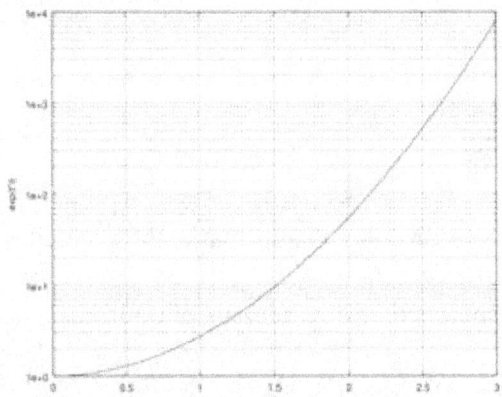

Figure 2.8   A semi-log plot

List 2.7
clear,clf,cla
t = .1:.1:3;
semilogy(t, exp(t.*t))
grid on
xlabel('t'); ylabel('exp(t*t)')

Plotting an implicit function

If a function is given implicitly, for example,

$$y^3 + \exp(y) = \tanh(x)$$

it cannot be expressed by x as a function of y, nor y as a function of x.

The function can be plotted, however, using **contour**. More detail of this approach is discussed in the subsection for contour of Section 2.3.

Plotting multiple curves

To plot two or more curves with a single **plot** command, write all pairs of coordinates repeatedly in the **plot** command:

```
List 2.8
Clear, clf, cla
x = 0:0.05:5;
y = sin(x);
z = cos(x);
plot(x,y,x,z)
```

Different line type or color is automatically selected for each curve by default. If desired, however, selected line color, line type, or mark, may be specified after each pair of coordinates; for example,

plot(x, y,'-g', x, z,':')

Octave allows to place '**linewidth**' for each data set to be plotted like

plot(x, y,'-g', 'linewidth', 1, x, z,':r', 'linewidth', 4)

On the other hand, Matlab despises the **linewidth** before the second data set. So plot each data set separately as

```
plot(x, y,'-g', 'linewidth', 1); hold on
plot(x, z,':r', 'linewidth', 4)
```

hold on, hold off

Until now we plotted all the curves at once with a single **plot** command. It often becomes desirable, however, to add a curve to a graph that has already been plotted. Such additional plotting can be done using **hold on** (see Figure 2.9 plotted by List 2.9).

```
List 2.9
clear,clf,cla
x = 0:0.05:5;
y = sin(x); plot(x,y); hold on
z = cos(x); plot(x,z,'--')
xlabel('x', 'fontsize',14); hold off
ylabel('y(-) and z(--)','fontsize',14)
```

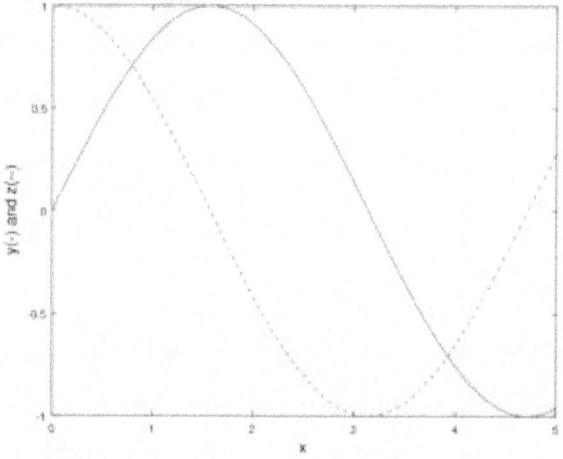

Figure 2.9   Two curves plotted with hold on

Once the **hold on** command is issued, the graph may stay on even when another script is executed. Therefore, it is prudent to place a **hold off** command at the end of the script.

73

When multiple curves are plotted with **hold on**, it is desirable to specify minimums and maximums of the coordinates on the graphic domain by the **axis** command. Otherwise, the limits are determined by default based on the first curve, which may cause other curves to be clipped.

The **hold on** command also becomes important when a time-consuming plot is undertaken for the following reason. The command to change parameters for figures such as **axis, colormap, view**, and other parameters can be used after a figure is plotted, but each time a new command is issued, the whole figure is re-plotted. In order to save time, give all the parameter com-mands before plotting, hold with **hold on**, and then use **plot**.

legend

The legend in a figure can be added by the **legend** command. The legend in Figure 2.10 is created by **legend('sin(x)', 'cos(x)')** added in List 2.10:

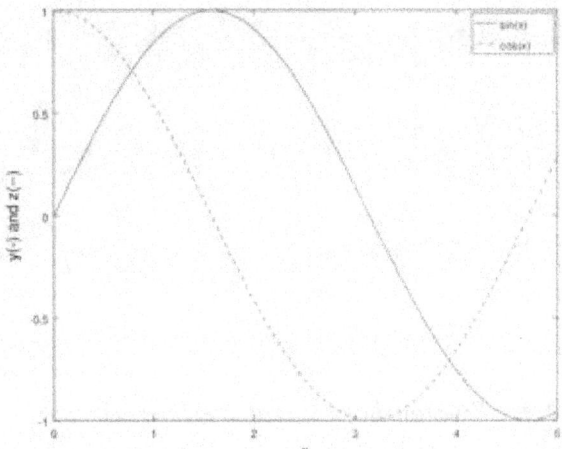

Figure 2.10   Illustration of the use of the **legend** command

List 2.10
clear,clf,cla
x = 0:0.05:5;

```
y = sin(x); plot(x,y); hold on
z = cos(x); plot(x,z,'--')
xlabel('x', 'fontsize',14);
ylabel('y(-) and z(--)','fontsize',14)
legend('sin(x)','cos(x)')
```

title

A title may be added to the top of the figure by the **title** com-mand.

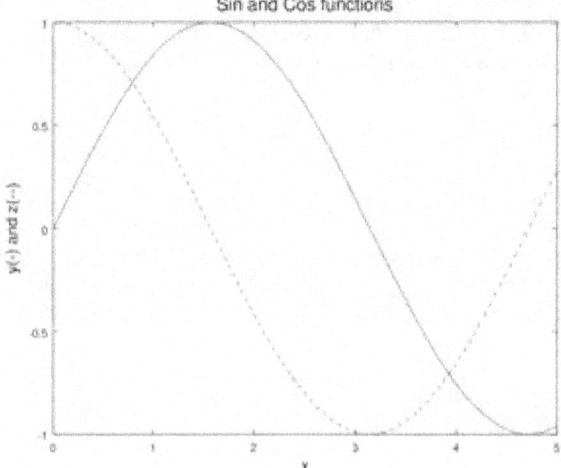

Figure 2.11   Figure with a title on the top

```
List 2.11
clear,clf,cla
x = 0:0.05:5;
y = sin(x); plot(x,y); hold on
z = cos(x); plot(x,z,'--')
xlabel('x', 'fontsize',14);
ylabel('y(-) and z(--)','fontsize',14)
title('sine and cosine functions','fontsize',16)
```

text

Text can be written in a graph by text. It needs three parameters in the argument, namely, **text(x,y,'string')**. The first two are x and y values of the location where the string starts. The third is a string

75

variable to be printed. The string variable can be a text enclosed by quote signs or a predefined string variable. For illustration, two **text** commands are used in List 2.12 to plot Figure 2.12.

```
List 2.12
clear,clf,cla
x = 0:0.05:5;
y = sin(x); plot(x,y); hold on
z = cos(x); plot(x,z,'--')
xlabel('x', 'fontsize',14);
ylabel('y(-) and z(--)','fontsize',14)
text( 1.65, 0, 'sin(x)' ,'fontsize',16)
text( 3.2, 0, 'cos(x)' ,'fontsize',16)
```

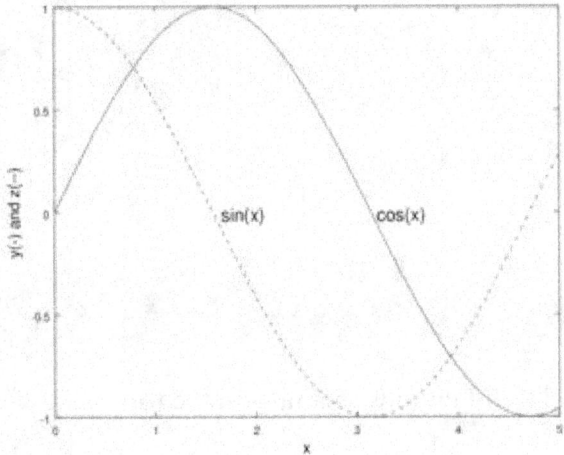

Figure 2.12   Illustration of the use of **text**

Text colors and other properties of text

Color and font size of a text in the graph may be changed. For example,

text(0.3, 0.2 ,'string','fontsize',18,'color','r')

will print string in red color with font size 18. If a default color is to be changed to green, for example, use:

set(gcf,'DefaultTextColor', 'g')

Thereafter, the text will be typed in green unless specified in each text command. Color for text may be chosen from red, yellow, green, cyan, blue, and magenta, which are abbreviated by 'r', 'y', 'g', 'c', 'b', and 'm', respectively. Color may be changed as many times as necessary.

Greek letters may be printed by **text**. For example,

    Figure
    axis([1.4 1.8   -0.1 0.1])
    text(1.5,0, '\alpha\beta\delta\epsilon\gamma\Gamma
    \Omega', 'fontsize',36)

will print Greek letters in a figure, as illustrated below:

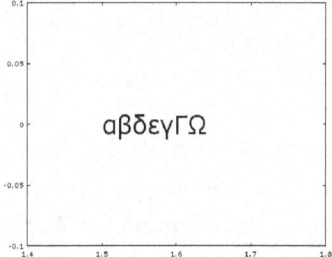

Other symbols may also be included, all according to the LaTex.

The font size of the axis tic mark values may be changed by the **set** command; for example,

    set(gca, 'fontsize',18)

changes the font of axis to size 18.

Superscript and subscripts may be written in a graph by **text**. To print a single character or a group of characters as superscript, use the caret symbol ^ followed by a single letter or group of letters enclosed by {}. For example, **text(2, 4, ' x^{-2} ')** will print

x^{-2} as text at x=2 and y=4. Subscript is the same as superscript except the ^ is replaced by underscore: for example, text(2, 4, ' s_{i,j}') will print $s_{i,j}$ as text. For illustration,

```
Figure
axis([1.4 1.8   -0.1 0.1])
text(1.5,0, 'y_{i,j} =    x^{2}', 'fontsize',36)
```

plots the following figure:

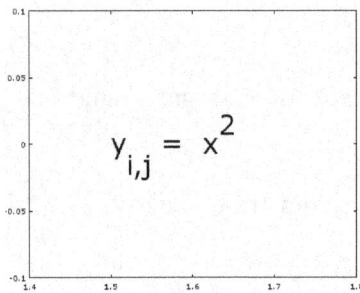

subplot

With **subplot(m,n,k)** command, mxn graphs are plotted in a single figure, where m, n, and k are integers. Here, the pair of m and n means a mxn array of graphs, and k is the sequential number of the graph. For example, **plot** after **subplot(3,2,1)** will plot the first graph in the 3x2 figures. The following script plots four graphs, as illustrated in Figure 2.13:

```
List 2.13
clear,clf,hold off
t=0:.3:30;
subplot(2,2,1),
    plot(t,sin(t)),title('SUBPLOT 2,2,1')
    xlabel('t'); ylabel('sin(t)')
subplot(2,2,2),
    plot(t,t.*sin(t)),title('SUBPLOT 2,2,2')
    xlabel('t'); ylabel('t.*sin(t)')
subplot(2,2,3),
    plot(t,t.*sin(t).^2),title('SUBPLOT 2,2,3')
```

```
    xlabel('t'); ylabel('t.*sin(t).^2')
subplot(2,2,4),
    plot(t,t.^2 .*sin(t).^2),title('SUBPLOT 2,2,4')
    xlabel('t'); ylabel('t.^2.*sin(t).^2')
```

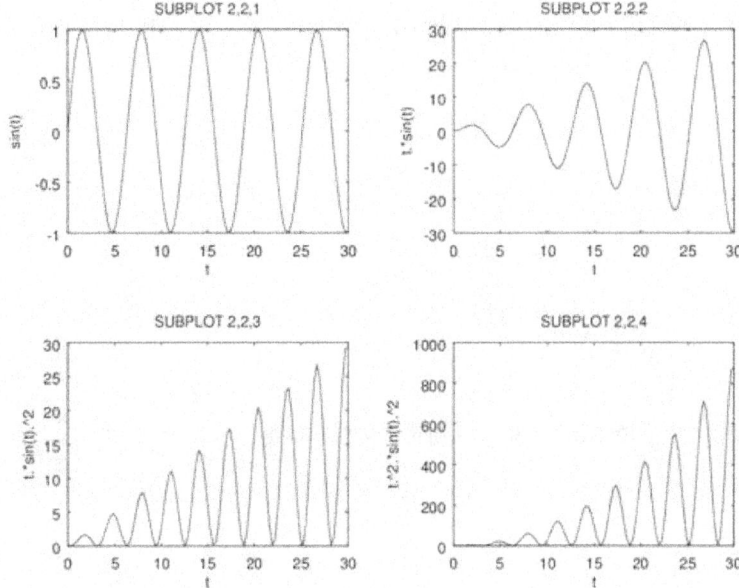

Figure 2.13   Plotting multiple figures by **subplot**

A vertical stack of two graphs is plotted by

```
subplot(2,1,1), plot(
subplot(2,1,2), plot(
```

Likewise, a row of two graphs is plotted by

```
subplot(1,2,1), plot(
subplot(1,2,2), plot(
```

3d plot

The command **plot3** is the three-dimensional version of **plot**. All the rules explained for **plot** apply to **plot3**. A spiral motion of a particle from point A to B in Figure 2.14 is plotted by List 2.14.

The view angle may be changed by **view**, as explained in more detail in Section 2.3. The axis command,

axis([x-min,x-max,y-min,y-max,z-min,z-max])

may be used to define bounds of the three-dimensional space.

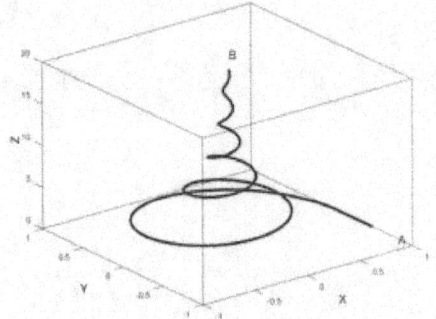

Figure 2.14   3D plot with **axis**

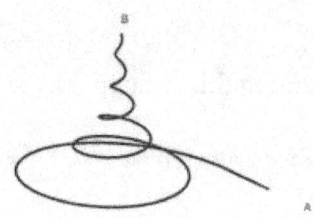

Figure 2.15   3D plot with **axis off**

List 2.14
clear,clf
t=0:0.1:20;
r= exp(-0.2*t);
th=pi*t*0.5;
z=t;
x=r.*cos(th);
y=r.*sin(th);
plot3(x,y,z, 'linewidth', 3)
hold on

```
plot3([1,1], [-0.5,0], [0,0], 'linewidth', 3)
text( 0.9,-0.95,1, 'A', 'fontsize',14)
n=length(x);
text( x(n),y(n),z(n)+2,'B', 'fontsize',14)
xlabel('X', 'fontsize',14);
ylabel('Y', 'fontsize',14);
zlabel('Z', 'fontsize',14);
```

2.2 How to Print or Record Graphs

print

The command to produce a graphic file of the current figure in jpeg form is **print filename.jpg** or **print -djpg filename**. To create a post script file of a figure, the command is print **filename.ps** or **print -dps filename**. See the next subsection in case the figure includes texts with subscripts or superscripts or Greek letters.

The figures in jpeg form may be inserted in the MS Word. The figure file in the post script form may used in LaTex editor.

Copy & paste

Figures can be copied from figure windows and pasted on MS Word by copy & paste operation using mouse.

Unfortunately, the figures that includes texts with subscripts/ superscripts or Greek letters do not come out well with the **print** command nor by copy & past. In this case, a screen capture technique is recommended. That is, click on the figure to copy. Push *Ctrl key* and *Alt key* and *Print Screen key* together. This action copies the entire figure window including its frame. Open the *Paint* software on PC. Paste on it by pushing *Cntl key* and *character v key* together. In the Paint software, you will see the figure, but it includes the window frame. The Paint software has a rubber band tool by which you can select the area of the figure excluding the figure window frame. Copy the selected area, and paste on MS Word. If you save the Word document in html format with a **filename**, the figure in jpg format becomes available in the file named **filename_files**.

Plotting by black only

There is a little secret in creating the 3-dimensional plots in this book. Because they are plotted in the figure window with color, some colors, particularly the colors near yellow, fade out because the book is printed by black only. Therefore it became necessary to plot only in black. This was done by typing the command

set(gcf, 'colormap', zeros(64,3))

on the key board after a color graph is plotted. This command changes the color map to all black whatever the original color is.

2.3   Plot of two-dimensional functions

Mesh plot

Suppose $x(i)$, $i=1,2, \ldots$ imax, are values on the x coordinate in increasing order, and $y(j)$, $j=1,2, \ldots$jmax, the values on the y coordinate in increasing order. Consider vertical lines passing through the points on the x-coordinate, namely, $x = x(i)$, and horizontal lines passing through the points on the y-coordinate, namely, $y = y(j)$, then the intersections between the two families of the lines form a mesh. A mesh point in this mesh may be denoted by the pair of indices (i,j) or pair of coordinates $(x(i), y(j))$. If we define a two-dimensional array of x as

$$x(i,j) = x(i)$$

and that of y as

$$y(i,j) = y(j)$$

then a function on the two-dimensional coordinate, $z = f(x,y)$, may be discretely represented by an array z:

$$z(i,j) = f(x(i,j), y(i,j))$$

In writing an m-file script, we denote the one-dimensional array of x(i) by x1, and the same of y(j) by y1; the two-dimensional array of x(i, j) by x2, and the same of y(i, j) by y2. Once the values of x1 and y1 are set, then two-dimensional arrays, x2 and y2, can be computed by the **meshgrid** command as

[x2, y2] = meshgrid(x1, y1)

We assume the function z = f(x, y) is given by

$$z = f(x,y) = x \exp(-x^2 - y^2), \qquad -2 < x < 2, \qquad -2 < y < 2$$

then, the two-dimensional array z is computed by

z = x.*exp(- x.^2 – y.^2)

A three-dimensional plot of z is done by the **mesh** command as

mesh(x, y, z)

A whole script is shown in List 2.15, which produces Figure 2.16.

```
List 2.15
clear, clf
x1 = -2:.2:2;
y1 = -2:.2:2;
[x2,y2] = meshgrid(x1,y1);
z2 = x2 .* exp(-x2.^2 - y2.^2);
mesh(x2,y2,z2)
title('This is a 3-D plot of   z = x * exp(-x^2 - y^2)')
xlabel('x'); ylabel('y'); zlabel('z');
```

In Figure 2.16, the three-dimensional plot is viewed from a default point of view eyes in space, but the location of the view eyes may be changed by the **view** command. In Figure 2.17, the view point is changed to [0.1, 1, 0.5], or equivalently x=0.1, y=1, and z=0.5.

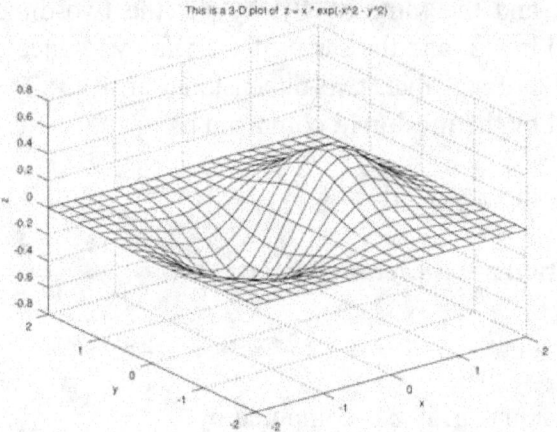

Figure 2.16   3D plot by `mesh`

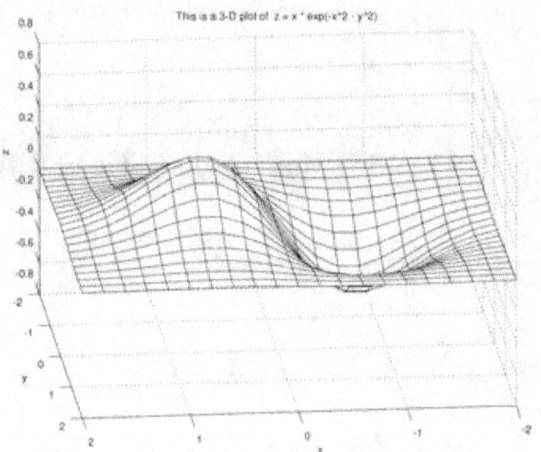

Figure 2.17   3D plot with a changed view

Figure 2.18 is a plane view of the mesh on the x-y plane plotted by **mesh**:

```
mesh(x2,y2,0*x2);
view([0,0,100001]);
xlabel('x'); ylabel('y')
```

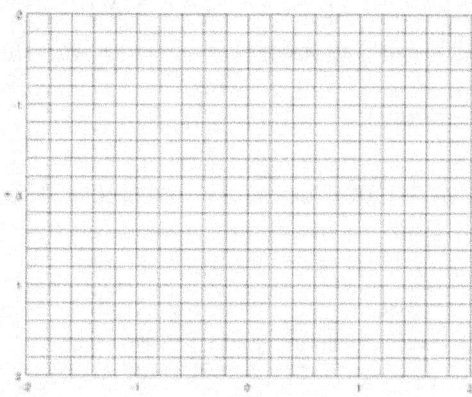

Figure 2.18   Plot of the mesh on the x-y plane

Contour plot

Contour is another way of visually expressing the distribution of a two-dimensional function.

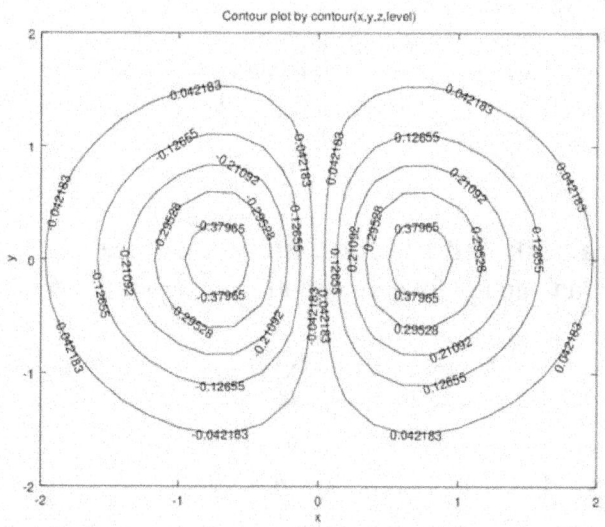

Figure 2.19   Contour plot of $z = x \exp(-x^2 - y^2)$

Assuming that the mesh grid data and functional data used in the prior section are utilized again, we simply write them in the command to plot contour:

contour(x, y, z, level)

where **level** is an array of contour levels, which are based on the interval between minimum and maximum values of z divided into a given number of subintervals. The number of levels is 10 in the present example. Contour levels are the mid points of the subintervals. Figure 2.19 of contour is plotted by List 2.16.

```
List 2.16
clear, clf
x1 = -2:.2:2;
y1 = -2:.2:2;
[x2,y2] = meshgrid(x1,y1);
z2 = x2 .* exp(-x2.^2 - y2.^2);
zmax=max(max(z2)); zmin=min(min(z2));
dz = (zmax-zmin)/10;
level = zmin + 0.5*dz: dz: zmax;
h=contour(x2,y2,z2,level); clabel(h)
title('Contour plot by contour(x,y,z,level)')
xlabel('x'); ylabel('y');
```

Plotting an implicit function using contour

Plotting of an implicit function mentioned earlier in Section 2.1

$$y^3 + \exp(y) = \tanh(x)$$

is now possible by **contour**. We rewrite the foregoing equation and define a two-dimensional function z(x, y) as

$$z(x, y) = y^3 + \exp(y) - \tanh(x)$$

We plot its contour for only one level of z=0. The plot of this function is illustrated in Figure 2.20, plotted by the script in List 2.17.

List 2.17
```
clear, clf, cla
xm = -3:0.2:3;    ym = -2:0.2:1;
[x, y] = meshgrid(xm, ym);
f = y.^3 + exp(y) - tanh(x);
contour(x,y,f,[0,0],'linewidth',2)
xlabel('x', 'fontsize',16);
ylabel('y', 'fontsize',16)
```

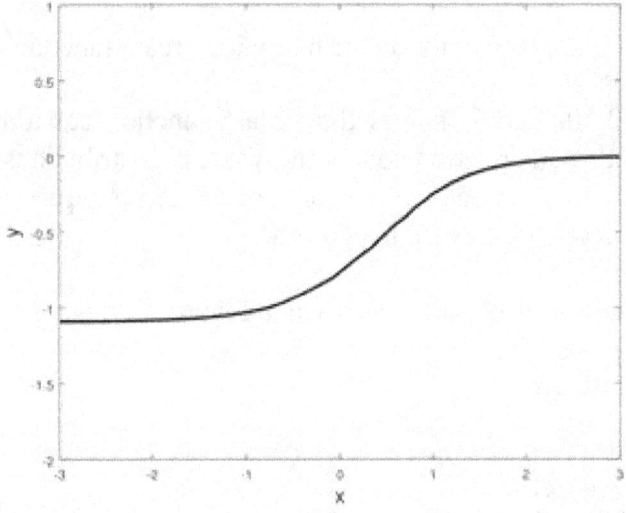

Figure 2.20   Plot of $z = y^3 + \exp(y) - \tanh(x)$

quiver

The **quiver** command is designed to plot 2D fluid velocity distributions and plots a vector indicating the magnitude and direction of the flow at each grid point. The **quiver** command requires the 2-dimensional arrays of x and y, plus u and v where u is the velocity component in the x-direction and v the velocity component in the y-direction. The format of this command is

quiver(x,y,u,v)

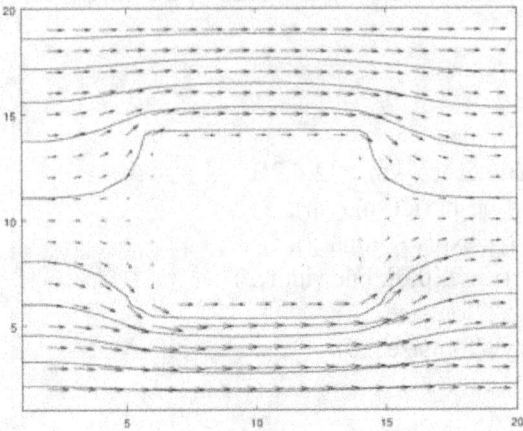

Figure 2.21 Velocity vector plot with stream function contour

In List 2.18, Part A solves the stream function equation for the inviscid flow, Part B computes the velocity distributions u and v, and in Part C, the velocity vectors are plotted by **quiver**, and the stream function is plotted by **contour**.

The results of plot are shown in Figure 2.21.

```
List 2.18
clear
clf
% PART A
imax=20;
jmax=20;
for i=1:imax
    for j=1:jmax
        x(i,j)=i;
        y(i,j)=j;
        f(i,j)=1;
        if j==1,    f(i,j)=jmax; end
        if j==jmax,    f(i,j)=1; end
    end
end
f(1:2,1:2)
```

```
i1=imax/2-5
i2=imax/2+5
j1=jmax/2-5
j2=jmax/2+5

for k=1:600
   for i=1:imax
      for j=2:jmax-1
         if i==1, f(i,j)=f(i+1,j); end %if
         if (i>1&i<imax),
            f(i,j)=0.25*( f(i-1,j)+f(i+1,j) + f(i,j-1)+   ...
            f(i,j+1))*1.9-0.9*f(i,j);
          end %if
         if (i==imax) f(i,j)=f(i-1,j);
         end %if
         if (i>i1 & i<i2 & j>j1 & j<j2),f(i,j)=jmax/2;
         end %if
      end
   end
end

% PART B
u(imax,jmax)=0;
v(imax,jmax)=0;
for i=2:imax-1
   for j=2:jmax-1
      u(i,j) = -(f(i,j+1)-f(i,j-1))/2;
      v(i,j) = (f(i+1,j)-f(i-1,j))/2;
   end
end

%PART C
quiver(x,y,u,v), hold on
contour(x,y,f)
```

2.4   Plotting of 3-dimensional structures

Octave has a number of graphic commands useful to develop images of 3D structures. Scripts used to plot the following figures are listed in http://octave.ismr.us/Commuter-Airplane.htm

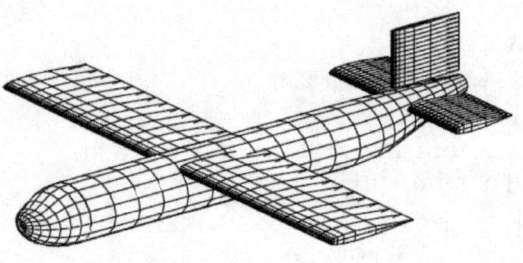

Plot 1: Plotted by mesh command

Plot 2: Plotted with surfl and shading interp

2.5 Bar charts

The graphs in this section are in color when produced by Octave/Matlab, which needed to be printed only in black/white in the book. However, the same graphs in color can be seen in http://octave.ismr.us/bar-color.htm.

With the bar(x) command, a bar chart is drawn, where x is an array of the data.

>> bar([1 3 2 1])

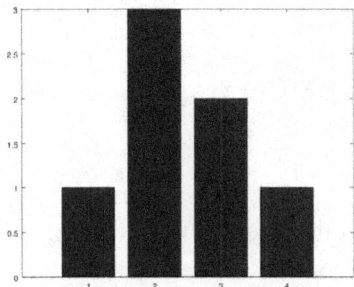

If x is a two-dimensional array of two rows, each column is considered as a category, and each row is considered to be a group such as a year or month.

>> bar([1 3 2 1; 2 1 1 3])

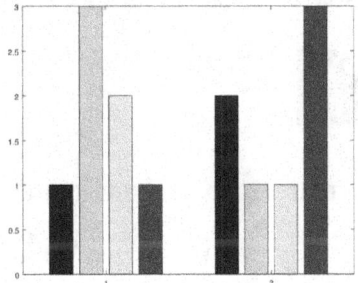

With the bar(x, "stacked"), the bars in each group are stacked.

>> bar([1 3 2 1; 2 1 1 3], "stacked")

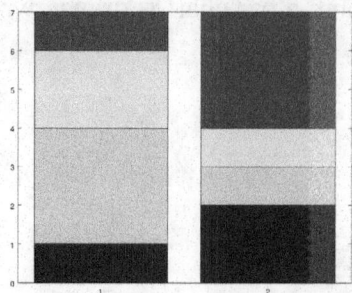

The width of bars are changed by w in bar(x,w,…). In the following example w=0.3 is used:

>> bar([1 3 2 1; 2 1 1 3], 0.3,"stacked")

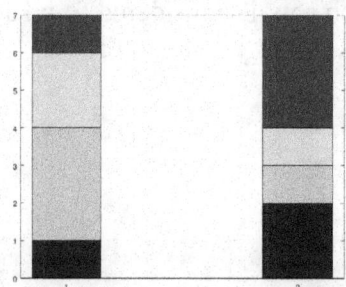

The bar(x, w, "hist") creates a bar chart in the histogram style:

>> bar([1 3 2 1 6 4; 2 1 1 0.5 3 5], 0.8,"hist")

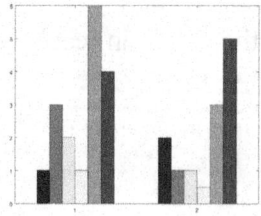

The bar(x, w,"histc") is the same as bar(x, w,"hist") except the histograms is left adjusted:

>> bar([1 3 2 1 6 4; 2 1 1 0.5 3 5], 0.8,"histc")

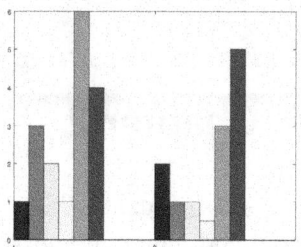

Legends may be added by the **legend** command:

```
>>bar([1 3 2 1 6 4; 2 1 1 0.5 3 5], 0.8,"histc")
>>axis off; legend('a', 'b', 'c', 'd', 'e', 'f')
```

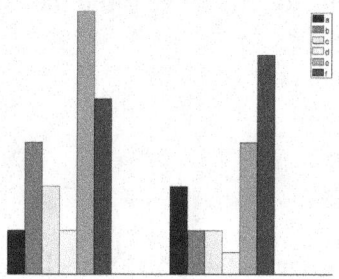

By removing the axis and adding texts, the meaning of each group can be printed under each group:

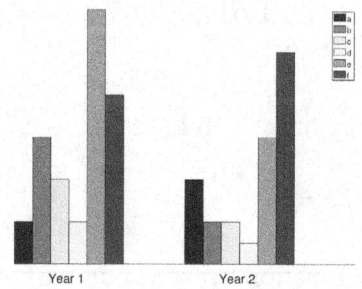

```
>>clf
>>bar([1 3 2 1 6 4; 2 1 1 0.5 3 5], 0.8,"histc")
>>axis([1 3 -0.5 6]); axis off;
>>legend('a', 'b', 'c', 'd', 'e', 'f')
>>text(1.2,-0.3, 'Year 1', 'fontsize', 16)
```

>>text(2.2,-0.3, 'Year 2', 'fontsize', 16)

Exercise problems for Chapter 2

[1] Plot the following functions in -1≤x≤1 using the color specified for each. The graph needs xlabel and ylabel in legible font size.
The curves must be plotted with linewidth 2.

  $y=\sin(x)\exp(-x)$ (green)
  $y=\cos(5a\cos(x))$ (blue)

[2] Plot the following functions together using the method of
plotting implicit functions for -1<x<3, -2<y<3.

$$2(x-1)^2 + (y-1)^2 = 3$$

$$(x^{1.5}-3)^2 + y^2 = 2$$

[3] Plot the following function with xlabel and ylabel:

$$y = 1/(1 + (x-2)^2), \quad 0 \le x \le 4$$

[4] Plot $y=\tan(x)$ in $0 \le x \le 10$. Does it look awesome or awk-
ward? Propose how to make it awesome.

[5] Plot the following equation in x>0 and y>0.

$$\sqrt{(1+x^2+\exp(y))} = 4\sin(xy)$$

[6] Repeat [1] and write the function near the curve of each as text.

[7] Repeat [1] and add the definition of the function as legends.

[8] Plot contour graph of the following function in x>0 and y>0
with contour lebels.

$$f(x,y) = 1/(1+ (x-1)^2 + (y-1)^2)$$

[9] Plot the following data in a bar graph

Year	Quantity a	b
1	25	12
2	31	13
3	29	17
4	40	19
5	35	21

Appendices

Appendix 1
Efficient ways of defining mathematical functions
(Use of eval and feval explained)

Most primitive way of using mathematical functions is to write the function directly in the main script. In this way, however, the same function may have to be written over and over if the function is used multiple times in the whole script. However, there are many convenient ways of defining and using it:

(1) Define a function as a separate function and save it as a m-file

Define a function m-file, for example

```
function f=my_f(x)
f=sin(2*x)+x.*2;
```

and saved in the same directory as the script calling it. Do not forget to use array arithmetic operators such as .^,   .* and ./ in the function m-file.

Then, it can be used in a script:

```
clear
x=0:0.1:3*pi;
f=my_f(x);
plot(x,f)
```

Alternatively, feval may be used:

```
clear
x=0:0.1:3*pi;
f=feval('my_f', x);
plot(x,f)
```

(2) A function name may be defined as string in the beginning of a program, and then **eval** is used whenever using the function is necessary

```
clear
x=0:0.1:3*pi;
s='my_f(x)';
f=eval(s);
plot(x,f)
```

(3) The function name may be read as input:

```
clear
x=0:0.1:3*pi;
%s='my_f(x)';
s=input(' Input your function name:   ')
f=eval(s);
plot(x,f)
```

Here, my_f must be enclosed by single quote signs as:

```
Input your function name:      'my_f'
s = my_f
```

By adding "s" at the end of **input command,** the input does not have to be enclosed by single quote signs:

```
clear
x=0:0.1:3*pi;
s=input(' Input your function name:   ',"s")
f=feval(s,x);
plot(x,f)
```

Input procedure:
```
Input your function name:   my_f
s = my_f
```

However, "s" works only in Octave, but not in Matlab.

(4) Methods of defining the function without function m-file

The definition of an equation can be inputted without using function m-file. In the following script, the equation is inputted through the keyboard (not enclosing by single quote signs):

```
clear
x=0:0.1:3*pi;
s=input(' Input your function name:   ',"s")
f=eval(s);
plot(x,f)
```

Input procedure:

```
Input your function name:   f=sin(2*x)+x.*2
s = f=sin(2*x)+x.*2
```

Here, s=f=… may be confusing, but s is the sting name and f=sin(2*x)+x.*2 is the string.

As already written, "s" works only in Octave, but not in Matlab.

Appendix 2
Examples of using eval command

In this appendix, item (4) in Appendix 1 is explained in more details. We show how to make writing a program that uses the same mathematical function many times easy. To explain how, consider the script below to solve the nonlinear equation,

$$\log(1+x) - x^2/10=0$$

by Newton's iteration (see Section 3.4):

```
% Appendix 2 Without using eval
clf,clear
x4plot=0:0.1:10; xinitial=5;
dx=0.001; x=xinitial;
for k=1:20
z=x; f=log(1+x)-x.^2/10; plot(x,f,'*'); hold on
xp=x+dx; fp=log(1+xp)-xp.^2/10;
fd=(fp-f)/dx; x=x-f/fd;; if abs(f/fd)<0.000001, break; end %if
```

```
end; residue=f/fd;
fprintf(['Equation="','log(1+z)-z.^2/10', '=0": Solution=%.5f\n'], x)
fprintf('Total iteration=%i, Residue=%.3e \n', k, residue)
x=x4plot; z=x; f=log(1+z)-z.^2/10;
plot(x,f)
```

The same equation is written several times in the script, so adapting it to another function needs much work. In order to make it easier to change, it is desired to write the equation only once in the entire script.

This is achieved by defining the function as a string variable, and using it by the **eval** command. In the following revised script, the function is written only once to define as string variable **fn**. Whenever the function is used, it is evaluated by **eval(fn)** with an appropriate definition of the argument just before **eval(fn)** is used. When the script is altered for another equation, only the second line needs to be changed, although the initial guess and **x4plot** that defines the plotting parameters may have to be adjusted also. If the revised script below is saved under a proper name, it is much more convenient to apply to many different equations than the previous one.

```
% Appendix 2 Using eval
clf,clear
fn='log(1+z)-z.^2/10';
x4plot='0:0.1:10'; xinitial=5;
dx=0.001; x=xinitial;
for k=1:20
z=x; f=eval(fn); plot(x,f,'*'); hold on
xp=x+dx;
fp=log(1+xp)-xp.^2/10;
z=xp; fp=eval(fn);
fd=(fp-f)/dx; x=x-f/fd; if abs(f/fd)<0.000001, break; end %if
end;
residue=f/fd;
fprintf(['Equation="',fn, '=0": Solution=%.5f\n'], x)
fprintf('Total iteration=%i, Residue=%.3e \n', k, residue)
x=eval(x4plot); z=x; f=eval(fn);
plot(x,f)
```

Appendix 3
How to make MyOctavePlace a default directory

Let us assume that Octave was downloaded and installed very recently, so the home directory where Octave opens by default is now

   c:/users/owner/

or an equivalent name because the name of directory **owner** may be different on each different computer. In this writing we assume **owner** is the current default directory name when Octave is installed. The reader must understand this and adjust the name if the current default directory name is different. We also assume that a directory **MyOctavePlace** was created in **c:/users/owner/** with an intention to make it Home directory for Octave works.

   We now wish to make

      c:/users/owner/MyOctavePlace

the Home directly so Octave will open automatically in this directory. Otherwise, working directory must be changed manually from **c:/users/owner/** to **c:/users/owner/MyOctave Place** by using **>>cd MyOctavePlace**.

   The following is the procedure in GUI Command Window. Copy the following two lines to the command window and run:

   >>cd c:/Octave/octave-4.0.0/share/octave/site/m/startup/
   >>edit octaverc

Then, the file octaverc opens in the Editor. Add the following line at the end:

   setenv('HOME', 'C:/users/owner/MyOctavePlace/'); cd ~/

Save the octaverc file. Close Octave and restart the computer.

If everything goes well with no errors, the Octave should open in MyOctavePlace directory as the Home directory. Make sure **>>pwd** is responded by

   c:/users/owner/MyOctavePlace

Appendix 4
Octave can become fatally sick

The readers should know **Octave can become fatally sick** if it swallows a wrong command or variable name. Here is the author's embarrassing experience.

At one time the author was writing a short m-file. During the first test, Octave stopped working properly, but there was no obvious error found in the m-file. However, whenever the command **plot** is used, error messages were generated. So the author tested the **plot** command independently of the m-file, but the result was the same. Other programs using **plot** stopped working in the same way.

Such ill behavior continued after Octave was shut down and re-opened, and even after the computer was shut down and rebooted. Another strange thing was that the m-file never ran even after almost all the contents including **plot** command were removed. It seemed as if Octave never forget that particular name of the m-file and continued attacking cantankerously.

Finally the only way for remedy was thought to re-install the whole Octave using the installer file downloaded.

It turned out later that the most probable cause was that the variable name **linewidth** was used accidentally as user-defined variable in the m-file tested. This name must be a reserved name used as a parameter in some graphic commands. My guess is that the "user-defined **linewidth**" permanently destroyed the Octave system, although **>>exist linewidth** returns **ans=0**.

After this incidence, though, Octave has been friendly to the author.

Matlab can possibly get similar symptom. The author has not attempted to cause the same trouble, nor is willing to try primarily because reinstallation of Matlab is not so simple to the author.

Solution of Problems

Exercise problems for Chapter 1

[1]

```
% Problem chapter 1 No.1
v=rand(1,100);
avarge=sum(v)/length(v);
n=0;m=0;
for i=1:length(v)
if v(i)>0.5, n=n+1; end
if v(i)<0.5, m=m+1; end
end %for
fprintf('Number of entries greater than 0.5 is %i \n',n)
fprintf('Number of entries less than 0.5 is %i \n',m)
```

[2]

```
% Problem Chapter 1 [2] (i)
clear all
 x=[-5, -2, 0, 3, 5];
for i=1:length(x)
 y(i) = x(i)^2 - 2*x(i) -2 ;
 fprintf('x= %.3f    y=%.3f \n', x(i), y(i))
end %for
```

```
x= -5.000    y=33.000
x= -2.000    y=6.000
x=  0.000    y=-2.000
x=  3.000    y=1.000
x=  5.000    y=13.000
```

```
% Problem Chapter 1 [2] (ii)
clear all
 x=[-5, -2, 0, 3, 5];
 y=x.^2 - 2*x -2 ;
for i=1:length(x)
fprintf('x= %.3f    y=%.3f \n', x(i), y(i))
end %for
```

[3]

```
% Problem Chapter 1 [3]
clear all
disp('Next input must be enclosed by single quotes');
```

```
s=input('Input the equation in a string: ');
disp('Next input must be placed in [   ] ');
x=input('Input the x values in an array: ');
eval(s)
for i=1:length(x)
fprintf('x= %.3f     y=%.3f \n', x(i), y(i))
end %for
```

Next input must be enclosed by single quotes
Input the equation in a string: 'y=x.^2 - 2*x -2'
Next input must be placed in [   ]
Input the x values in an array: [-5, -2, 0, 3, 5]
y =

    33   6   -2   1   13

x= -5.000   y=33.000
x= -2.000   y=6.000
x= 0.000   y=-2.000
x= 3.000   y=1.000
x= 5.000   y=13.000

[4]

```
%Problem Chapter 1 [4]
x=1;
while x<999
x=input('Input x value: ');
if x>=999, return
elseif x<0;
y=exp(x);
fprintf('x = %f, y=exp(x)= %f \n', x, y);
elseif x>0
y=1/(1+x^2);
fprintf('x = %f, y=1/(1+x^2)= %f \n', x, y);
end %if
end %while
```

Input x value: 9
x = 9.000000, y=1/(1+x^2)= 0.012195
Input x value: 0
Input x value: -9
x = -9.000000, y=exp(x)= 0.000123
Input x value: 999

[5]

```
% Problem Chapter 1 [5]
```

```
for N=[1000, 10000,100000];
v=rand(1,N);
count=0*(1:5);
for m=1:length(v)
bin=floor(v(m)/0.2)+1;
count(bin)=count(bin)+1;
end
f=count/N;
average=sum(f)/5;
variance=sum((f-average).^2)/5;
disp('  ')
fprintf('Number of random numbers N = %i\n', N)
disp('Fraction of N in each bin')
disp(f)
fprintf('Variance = %.3e\n', variance)
end %for
```

Number of random numbers N = 1000
Fraction of N in each bin
    0.19300   0.21800   0.18500   0.20400   0.20000
Variance = 1.228e-004

Number of random numbers N = 10000
Fraction of N in each bin
    0.19680   0.20180   0.20220   0.19620   0.20300
Variance = 8.352e-006

Number of random numbers N = 100000
Fraction of N in each bin
    0.19839   0.20184   0.20124   0.19974   0.19879
Variance = 1.809e-006

[6]

```
>> clear
>> a=hilb(5)
a =
    1.00000    0.50000    0.33333    0.25000    0.20000
    0.50000    0.33333    0.25000    0.20000    0.16667
    0.33333    0.25000    0.20000    0.16667    0.14286
    0.25000    0.20000    0.16667    0.14286    0.12500
    0.20000    0.16667    0.14286    0.12500    0.11111
>> a=a+1
>> save a
>> clear
>> load a
```

```
>> a
a =
    2.0000    1.5000    1.3333    1.2500    1.2000
    1.5000    1.3333    1.2500    1.2000    1.1667
    1.3333    1.2500    1.2000    1.1667    1.1429
    1.2500    1.2000    1.1667    1.1429    1.1250
    1.2000    1.1667    1.1429    1.1250    1.1111
```

[7]

```
u + v
    2    3    4    5
    3    4    5    6
    4    5    6    7
    5    6    7    8
c*v
    5
    8
    8
   16
c*w    Invalid operation
a+b
    2    4    4
    4    0    2
    5    1    4
b*c Invalid operation
a*b
   21    7   11
    4    0    3
    9    2    5
a.*b
    1    0    3
    4   -1    1
    4    0    4
u.^2
    1    4    9   16
```

[8]

```
>> a*b'    :Row vector times column vector becomes a scalar.
            Valid in linear algebra.
ans =   32
```

```
>>a'*b        :Column vector times row vector becomes a matrix.
              Valid in linear algebra.
ans =
      4     5     6
      8    10    12
     12    15    18
>> a.*b    :Array arithmetic operation. Not in linear algebra.
ans =
      4    10    18
>> a.^b    :Array arithmetic operation. Not in linear algebra.
ans =
      1    32   729
>> a.*b'   :Array arithmetic operation. Not explained in text,
              but makes sense.
ans =
      4     8    12
      5    10    15
      6    12    18
```

[9]

```
d=1:29;                             February Calendar
d=[0,d];n=0;                         0   1   2   3   4   5   6
disp(' February Calendar')           7   8   9  10  11  12  13
for k=1:ceil(d(length(d))/7);       14  15  16  17  18  19  20
for j=1:7                           21  22  23  24  25  26  27
n=n+1;                              28  29
if n>length(d); break; end%if
if d(n)<10,
fprintf('   %i', d(n));
else
fprintf(' %i', d(n));
end%if
end%for
fprintf(' \n');
end%for
disp('         ')
```

[10]

n	x	x^2	%Problem Chapter 1 [10]
1	0.0	0.000	n=(1:10)';
2	0.5	0.250	x=[0 0.5 1.1 1.6 2.2 2.7 3.4 3.9 4.4
3	1.1	1.210	5.0]';
4	1.6	2.560	xsq=x.^2;
5	2.2	4.840	fprintf(' n x x^2 \n')

6	2.7	7.290	for m=1:length(n)
7	3.4	11.560	if xsq(m)<10,
8	3.9	15.210	fprintf(' %i %.1f %.3f\n', n(m), x(m),
9	4.4	19.360	xsq(m))
10	5.0	25.000	else

```
if m<10
fprintf(' %i       %.1f        %.3f\n', n(m), x(m),
xsq(m))
end%if
if m>=10
fprintf('%i       %.1f        %.3f\n', n(m), x(m),
xsq(m))
end%if
end%if
end%for
```

[11]

```
%Chapter 1 [11]          Result:
A='Jopnes      ';        Z =
B='Smith       ';        Jopnes
C='Alexander';           Smith
D='Xing        ';        Alexander
E='Camden      ';        Xing
Z=[A;B;C;D;E]            Camden
```

Exercise problems for Chapter 2

[1]

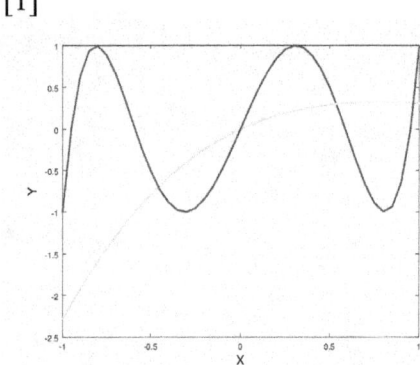

```
%Problem Chapter 2 [1]
x=-1:0.05:1;
y1=sin(x).*exp(-x);
y2=cos(5*acos(x));
plot(x,y1,'g', 'linewidth', 2);
hold on
plot(x,y2,'b', 'linewidth', 2);
xlabel('X', 'fontsize', 16)
ylabel('Y', 'fontsize', 16)
```

[2]

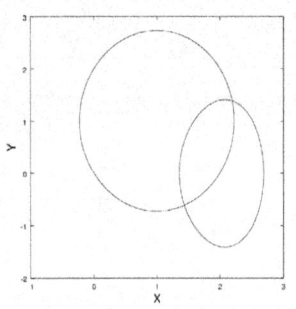

```
%Problem Chapter 2 [2]
clear all, clf
x1=-1:0.1:3;
y1=-2:0.1:3;
[x,y]=meshgrid(x1,y1);
f=2*(x-1).^2 + (y-1).^2 - 3;
contour(x,y,f, [0 0], 'k'); hold on
axis square
f=(x.^1.5-3).^2 + y.^2 -2;
contour(x,y,f, [0 0], 'k');
xlabel('X', 'fontsize', 16)
ylabel('Y', 'fontsize', 16)
```

[3]

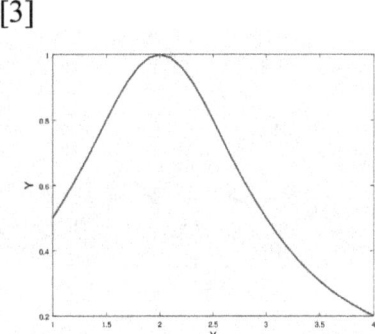

```
%Problem Chapter 2 [3]
clear all, clf
x=1:0.1:4;
y = 1./( 1 + (x-2).^2);
plot(x,y,'linewidth',2)
xlabel('X', 'fontsize', 16)
ylabel('Y', 'fontsize', 16)
```

[4]

Awful

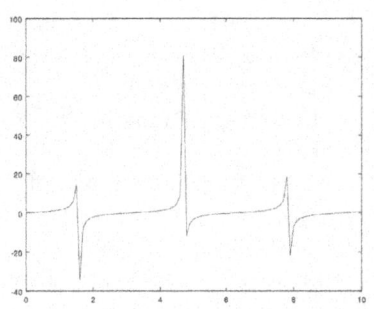

```
x=0:0.1:10;
plot(x,tan(x))
```

Awesome

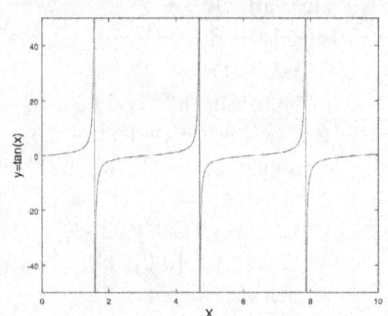

```
x=0:0.01:10;
plot(x,tan(x))
axis([0 10 -50 50])
xlabel('X', 'fontsize',
16)
ylabel('y=tan(x)',
'fontsize', 16)
```

[5]

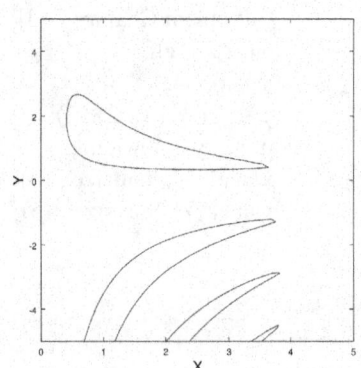

```
%Problem Chapter 2 [5]
clear all, clf
x1=0:0.1:5;
y1=-5:0.1:5;
[x,y]=meshgrid(x1,y1);
f=sqrt(1+x.^2+exp(y))-4*sin(x.*y);
contour(x,y,f, [0 0], 'k');
axis square
xlabel('X', 'fontsize', 16)
ylabel('Y', 'fontsize', 16)
```

[6]

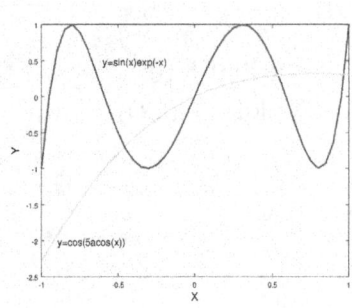

```
%Problem Chapter 2 [6]
clear, clf
x=-1:0.05:1;
y1=sin(x).*exp(-x);
y2=cos(5*acos(x));
plot(x,y1,'g', 'linewidth',2); hold
on
plot(x,y2,'b', 'linewidth',2);
xlabel('X', 'fontsize', 16)
ylabel('Y', 'fontsize', 16)
text(-0.6, 0.5, 'y=sin(x)exp(-
x)','fontsize', 14)
text(-0.9, -2,
'y=cos(5acos(x))','fontsize', 14)
```

[7]

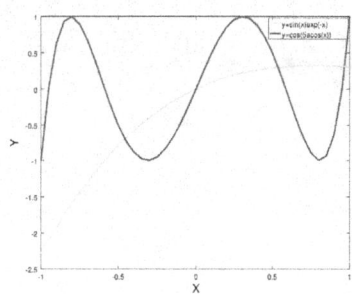

```
%Problem Chapter 2 [7]
clear, clf
x=-1:0.05:1;
y1=sin(x).*exp(-x);
y2=cos(5*acos(x));
plot(x,y1,'g', 'linewidth',2); hold on
plot(x,y2,'b', 'linewidth',2);
xlabel('X', 'fontsize', 16)
ylabel('Y', 'fontsize', 16)
legend( 'y=sin(x)exp(-x)',
'y=cos(5acos(x))' )
```

[8]

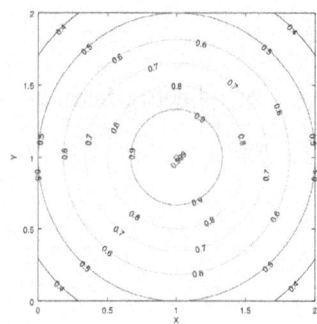

```
%Problem Chapter 2 [8]
x=0:0.05:2;
y=x;
[X,Y]=meshgrid(x,y);
f=1./( 1 + (X-1).^2 + (Y-1).^2);
level=[0.2:0.1:0.9,0.999];
h=contour(X,Y,f, level);
clabel(h)
xlabel('X')
ylabel('Y')
axis square
```

[9]

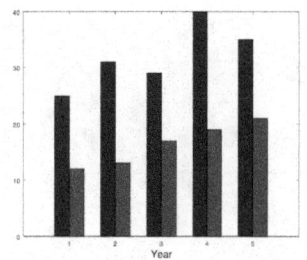

```
%Problem Chapter 2 [9]
data=[
1        25        12;
2        31        13;
3        29        17;
4        40        19;
5        35        21
]
```

```
bar(data(:, 2:3), 0.8, "hist")
xlabel('Year','fontsize',16)
```

References

S. Nakamura, Numerical Analysis and Graphic Visualization with MATLAB, 2nd ed., Prentice-Hall, 2002

P.J.G. Long, Introduction to Octave, http://www-mdp.eng.cam.ac.uk/web/CD/engapps/octave/octavetut.pdf, 2005

Octave vs Matlab Image Processing, http://wizzcore.com/tag/octave-vs-matlab-image-processing

J. S. Hansen, GNU Octave Beginner's Guide, Packt Publishing, 2011

J. W. Eanon, GNU Octave Manual, Network Theory Limited, 2nd ed., 2005

www.ingramcontent.com/pod-product-compliance
Lightning Source LLC
Chambersburg PA
CBHW060350190526
45169CB00002B/552